Scientists Must Write

To be effective as a scientist or engineer, or as a student, you must write well. This book, by a scientist, will help you to write: to observe, remember, think and plan; to manage your time and avoid stress; and to improve your communication skills.

Scientists Must Write is about the importance of writing in science and engineering, and the characteristics of scientific writing (Chapters 1 to 4); how to write and your choice and use of words (Chapters 5 to 8); using numbers and illustrations to support your writing (Chapters 9 and 10); finding information and citing sources (Chapter 11); writing project reports, theses and papers for publication (Chapters 12 and 13); and giving a short talk or presentation (Chapter 14).

This new edition retains the features that contributed to the success of the first edition, and has been updated to take account of the use of computers in recording, storing and processing data, information retrieval, and word processing. It also includes new appendices on punctuation, spelling and computer appreciation.

Robert Barrass is Emeritus Research Scientist at the University of Sunderland and has many years' experience of helping degree and diploma students to improve their writing and other key skills. His best selling books Students Must Write and Study!, like Scientists Must Write, are published by Routledge – as is his new book Writing at Work.

Also by Robert Barrass

Science
An introduction

Students Must Write
A guide to better writing in course work and examinations

Study!
A guide to effective learning, revision and examination techniques

Writing at Work
A guide to better writing in administration, business and management

Scientists Must Write

A guide to better writing for scientists, engineers and students

Second edition

Robert Barrass

London and New York

First published 2002
by Routledge
11 New Fetter Lane, London EC4P 4EE

Simultaneously published in the USA and Canada
by Routledge
29 West 35th Street, New York, NY 10001

Reprinted 2003

Routledge is an imprint of the Taylor & Francis Group

© 2002 Robert Barrass

Typeset in Goudy by M Rules
Printed and bound in Great Britain by
TJ International Ltd, Padstow, Cornwall

British Library Cataloguing in Publication Data
A catalogue record for this book is available from the British
Library

Library of Congress Cataloging in Publication Data
A catalog record has been requested

ISBN 0–415–26996–2

To Ann

Contents

Preface

Some people say that young scientists and engineers should be taught to write well so that they can be employed in administration, business and management. This is true but they must also be able to write good English if they are to be effective as scientists and engineers. The essentials in scientific and technical writing are the same: accuracy, clarity, completeness, simplicity, etc. (see Chapter 4) so, to avoid undue repetition, in this book the word *scientist* means scientist and technologist, and *scientific writing* means scientific and technical writing.

Writing is part of science, but many scientists receive no formal training in the art of writing. There is a certain irony in this: we teach scientists and engineers to use instruments and techniques many of which they will never use in their working lives, and yet do not teach the one thing they must do every day – as students, and in any career based on their studies.

This book, by a scientist, is not a textbook of English grammar. Nor is it just one more book on how to write a technical report, or a thesis, or a paper for publication. It is about all the ways in which writing is important to students and working scientists and engineers in helping them to observe, to remember, to organise, to plan, to think and to communicate.

As a guide to better writing, it is not intended to be read at one sitting. The early chapters should encourage young scientists and engineers to appreciate how important their writing is, every day, and help them to improve their written work. The later chapters should help them most when they have to prepare longer documents, as students undertaking project work or as employees preparing progress reports or papers for publication. Chapters 1, 2 and 3 are about all the ways in which writing is important in science, and Chapter 4 is about the characteristics of scientific writing. I hope Chapters 5 to 8 will help all those who have difficulty in putting their thoughts into words, and cause them to consider the words they use and how they use them. In scientific writing, numbers (Chapter 9) and illustrations (Chapter 10) are important, and the preparation of tables and illustrations is usually the first step in writing the Results section of a project report, thesis, or scientific paper for publication (Chapters 12 and 13). A chapter on information retrieval is included (Chapter 11) and one on talks and presentations (Chapter 14).

Much has changed in the twenty-five years since I wrote the first edition, published in 1978, which has never been out of print. The features that contributed to its success have been retained in this new edition, so those who know the book will find much that is familiar in both content and approach. However, changes have been made, where necessary, in all chapters – in the light of my experience in helping students to improve their written work, in placing students with different employers as part of our degree and diploma courses, and in visiting them and their supervisors in a variety of training laboratories. I have also made changes, at appropriate points in the text, to take account of the use of computers in the analysis of data and the presentation of results; the storage, communication and retrieval of information; and the preparation of documents. Extra material has also been added, for ease of reference, as three appendices: A: Punctuation, B: Spelling, and C: Computer appreciation.

Scientists Must Write may be used *either* as an alternative to a formal course on scientific and technical writing *or* to complement such a course. To help those who require guidance on a particular topic, a detailed list of Contents is included – and an Index. To help all readers, and to reduce the number of cross-references, some essential points are repeated in different contexts.

Exercises headed *Improve your writing* at the end of most chapters (and listed as *exercises* in the *Index*) are suitable for self-instruction but, where appropriate, suggestions are included to help teachers of science or of scientific writing to use these or similar exercises in their own courses. Examples of unscientific writing and/or poor English, included in some chapters, are accompanied by notes of faults or suggested improvements. Like Gowers (1986), I do not give the source of such extracts but they were written by people who speak English as their first language: some by professors in universities, and all by authors of books or contributors to journals.

Acknowledgements

I thank Jonathan Barrass for help in preparing this second edition, especially with the parts on aspects of information technology. I also thank Elizabeth Cunningham, independent IT Trainer and Consultant for reading the typescript for Appendix C, and colleagues in the University of Sunderland: Library staff for help with information retrieval; Paul Griffin and Richard Hall of the School of Sciences for their interest and for advice on the use of personal computers and on health and safety, respectively; and Christine Hillam of the School of Computing and Engineering Technology for reading and commenting on the whole book in typescript. The cartoons are by Dave Douglas. I also thank Ann, my wife, for her interest, help and encouragement.

Robert Barrass
Sunderland
August 2001

1 Scientists must write

When asked why we must write, most scientists and engineers think first of the need to communicate. Communication is so important in science and engineering that it is easy to forget our other reasons for writing. We write as part of our day-to-day work: to help us to observe, to remember, to think, to plan and to organise, as well as to communicate. Above all, writing helps us to think and to express our thoughts – and anyone who writes badly is handicapped both when working alone and in dealing with others.

When we write to people we know, they can judge us by everything they know about us – by our writing and by our conversation, appearance and behaviour. But when we deal with people whom we have never met, they judge us in the only way they can: by the way we communicate – on the telephone and in writing. For example, an application for employment that is well presented, informative and clearly expressed, in grammatically correct English, will create an immediately favourable impression of the writer as being well educated, well organised and clear thinking.

Only by writing well can we give a good account of ourselves as students (writing in course work and examinations), as applicants for employment, and as employees (writing, for example, letters, memoranda, instructions, progress reports, articles and reviews, and papers for publication).

Some scientists and engineers recognise the importance of writing in their work. They take trouble with their writing, and write well. Some, because they are satisfied with their writing, write without considering the possibility of improvement. They, and others who know that they write badly, are mistaken if they believe that writing is not particularly important in science and engineering.

Many people must be encouraged in their belief that their writing is satisfactory by their success in school and college examinations, but most students would get higher marks in course work and in examinations if they were better able to put their thoughts into words. Only teachers and examiners know how many marks are lost by students who do not show clearly whether or not they understand their work. In schools, many of the most able students fail to show their ability; and so do many in colleges and universities. They need help with their writing more than further instruction in their chosen subjects.

The need for improvement is also demonstrated in the writing of working scientists and engineers, who presumably try to write well when preparing accounts of their work for publication. Yet many submit verbose and poorly expressed compositions, containing errors in punctuation and grammar, which editors must return for correction and major revision or for rewriting before they can reconsider them for publication. Furthermore, despite the efforts of editors, many published papers include verbose and ambiguous sentences which indicate that many educated people either do not think sufficiently about what they write or are unable to express their thoughts clearly and concisely (for examples, see *Criticise other people's writing*, page 35, and *Edit the work of others*, page 76).

All scientists and engineers should accept that writing is part of their work, but the biggest difficulty facing anyone who would like to see an improvement in the general standard of scientific and technical writing is that most educated people are content with their writing. Indeed, although young scientists and engineers know that the study of mathematics is essential in providing a basis for their work, many may feel that English is more important for students of the arts and humanities. In fact, scientists and engineers spend much of their working lives writing with a pen or using a computer for word processing. Although arts students may have little need of mathematics, writing is important for all students and in all professions.

Unless scientists and engineers express their thoughts unambiguously, they will prepare records that, later, they cannot understand themselves – and write communications that others misinterpret. It does not matter in a novel if the author's meaning is not absolutely clear; indeed, much may be left to the reader's imagination. But scientists and engineers must express their thoughts clearly and simply – so that they cannot be misunderstood – because misunderstandings may cause other people wasted effort and result in costly mistakes, accidents and even loss of life.

Writing as part of science

The scientific method

Scientific research begins with a problem which may come from personal observation or from a consideration of work done by others. Problems are tackled by the method of investigation, in an attempt to obtain evidence related to a hypothesis. If the problem is stated as a question, then each hypothesis is a possible answer to the question or a possible explanation. The observations and measurements recorded during an investigation are data, and these are analysed and the results of analysis considered, and compared with results from other investigations. This leads to the bringing together of information from different sources, to synthesis, to the recognition of order (to classification), and to the making of generalisations (stated as norms, concepts, principles, theories and laws).

As one hypothesis is supported by new evidence and others are rejected, additional hypotheses may provide other possible explanations. Each hypothesis can be retained only for as long as it provides a satisfactory explanation for the observations accumulated on the subject. When a hypothesis is generally accepted by scientists working in the field it may be called a theory, and may lead to the statement of a principle or law which has value not only because it accounts for observations which have been made, but also because it allows them to predict what will happen in future observations and experiments.

Communication is involved at all stages in the application of the scientific method. The hypothesis upon which each investigation is based may come from personal observations, but each scientist should know of the observations and experiments of other scientists who are working on the same problem or in the same area of study. This helps to prevent unnecessary duplication of effort (but see page 117), and should also result in a contribution to knowledge by ensuring that new observations are related to what is already known.

Hypotheses, theories and laws must be modified or discarded if at any time they are found wanting, or if a better explanation is suggested for the accumulated observations on the subject. Even if scientists work alone, therefore, the scientific method makes science a co-operative venture and no work is complete until a report has been written.

The publication of research

The literature of science, a permanent record of the communication between scientists, is also the history of science: a record of the search for truth, of observations and opinions, of hypotheses that have been ignored or have been found wanting or have withstood the test of further observation and experiment. Science is a continuing endeavour in which the end of one investigation may be the starting point for another. *Scientists must write*, therefore, so that their discoveries may be known to others.

> Their purpose is, in short, to make faithful *Records*, of all the Works of *Nature*, or *Art*, which can come within their reach: that so the present

Age, and posterity, may be able to put a mark on the Errors, which have been strengthened by long prescription: to restore the Truths, that have lain neglected: to push on those, which are already known, to more various uses: and to make the way more passable, to what remains unreveal'd. This is the compass of their Design.

Thomas Sprat (1667) *History of The Royal Society*

The popularisation of science

Scientists must write formal accounts of their work for publication in journals which are read only by specialists, but which are accessible to scientists everywhere. Yet science is shaping our world, and whether they are pursuing knowledge for its own sake, or trying to solve practical problems, scientists must also write articles, reviews and books – about what they are doing and why, and about what other scientists are doing – for scientists working in other fields, for students of science, and for other interested people.

If we do not trouble to tell other people about science, or to discuss the impact of science on society, we should not be surprised if science and technology remain a closed book to many educated people, if the scientist is distrusted, if people do not appreciate the interdependence of pure and applied science, or if people expect too much of science.

In textbooks, scientists present science not only to tomorrow's scientists, but also to those who will work and take decisions in other fields. The writer of textbooks, therefore, has a unique opportunity to interest and inform. If people do not take an interest in science while they are young, they are unlikely to do so later.

Writing good books for young people is one of the most important duties that each generation of scientists must perform. The younger the age-group the scientists write for the more important is their work, because young children are quick to decide which subjects are of interest and which are not. If they do not understand or are not interested by their first books on any subject, the opportunity to capture their interest may have been lost.

Developing essential skills

It is not enough to teach science to young scientists and engineering to young engineers; they must also be helped to develop the skills needed in both study and employment (see Table 1.1).

There is a certain irony in teaching students of science and engineering to use techniques and instruments that they may never use in their working

Table 1.1 Key skills needed in study and in any career as a scientist or engineer, manager or administrator *

Personal skills	Some reasons for underachievement
1 Self management	Not prioritising tasks. Not establishing and working towards achievable long-, medium- and short-term objectives. Personal problems. Problems in dealing with others
2 Money management	Budgeting problems/Worries about money Lack of financial planning
3 Time management	Poor organisation: ineffective use of time Lack of foresight
4 Summarising	Inability to get to the heart of the matter
5 Information retrieval	Not filing notes for quick and easy access Not making good use of private records, personal contacts, and other sources of information
6 Processing information	Not bringing together relevant material from different sources in personal records and communications
7 Problem solving	Not thinking through issues to a satisfactory conclusion
8 Thinking and creativity	Not thinking critically. Mindless repetition of other people's thoughts: unwillingness to consider new approaches to problems or different points of view
9 Communicating	Not expressing thoughts clearly, concisely and convincingly when speaking and when writing

Note
* Key skills are sometimes called common skills or transferable skills because they are needed in studying all academic subjects and in all careers based on such studies. They are also called enterprise skills because they are characteristic of all those who show a readiness to take control of their lives.

lives, and yet not ensuring that they can express their thoughts clearly and simply in writing. This is something they will need to do every day as working scientists and engineers – and if they are promoted to management, they will spend more and more of their time communicating their thoughts in writing, in conversations on the telephone, and when speaking in meetings.

Those young people who, after qualifying as scientists and engineers, go directly into administration or start training for management will find that if they take trouble to improve their writing, they will produce work that is easier to read and understand. They will be more likely to consider the needs and feelings of others, and so they will be more effective as administrators or managers.

Teachers of science subjects in secondary, further and higher education can help to teach English, if they write well themselves, by telling young scientists why they need to write. Their students will not appreciate the importance of writing in all their school work if the teacher of English is the only one who corrects errors in spelling, punctuation and grammar. And they will not know that the requirements in scientific and technical writing are quite different from the imaginative writing expected in some English essays.

Young scientists and engineers should understand, as early as possible in their studies, that if they write well, they will be better students, achieve higher grades in course work and examinations, and be more effective in their careers.

The power of rightly chosen words is great, whether these words are intended to inform, to entertain, or to move. But there is no short cut to better writing: we learn most by trying to express our thoughts in writing, as clearly and simply as we can, by considering the comments of our teachers and colleagues or the advice of editors, and by example – by reading good prose.

Improve your writing

The exercises under this heading, at the end of most chapters, may be undertaken by readers working alone or used by tutors as ideas for incorporation in their courses on scientific and technical writing.

What scientists and engineers write

As a basis for discussion, tutors may start a course on scientific or technical writing by asking their students to prepare a list of the kinds of writing undertaken by students and by working scientists and engineers.

As a class exercise, students can be given ten to fifteen minutes – working alone – to prepare their separate lists. Then the tutor can take suggestions from the class and list them on a board or flip chart for consideration by the whole class. This may take another ten or fifteen minutes; and it may be necessary to remind the class that many of the things written by non-scientists are also written by scientists and engineers. The list can then be reconsidered by students as they answer the question: 'How does writing help scientists and engineers with their work?' Students should keep the notes made in this class for reconsideration later in their course.

Consider any comments on your written work

As a student you should appreciate and consider carefully each of the comments written on any assessed course work. But you are also fortunate if, as a working scientist or engineer, you have colleagues who write well and are willing to read and comment on any important documents you prepare. If possible, teachers of English, as well as teachers of scientific and technical writing, should base their instruction on constructive criticism of their students' own practical notebooks, reports and essays; and tutors providing in-service instruction should use communications produced in their own organisation. This is the best way, and perhaps the only way, to demonstrate to busy people that if they can improve their writing, they will be more effective in their work.

Write a set of instructions

A scientist or engineer, like anyone else, has to use instructions – and many accidents are the result of failures in communication attributable to ambiguous, incomplete or otherwise misleading instructions. Consider the kinds of instructions you give, and those you use. Are they all good instructions?

What do you expect of a set of instructions? What faults in a set of instructions may annoy the user, cause accidents, or result in other costly mistakes? Unless you have already given much thought to this subject, you will find it helpful to prepare a set of instructions headed *How to write a set of instructions*.

This is an exercise any reader can undertake alone, but it is more interesting as a class exercise, first with each participant working alone, for about ten minutes, then sharing ideas with someone else for another ten minutes, and then working in small groups as small committees trying to reach agreement.

Finally, and preferably when participants have had more time to think about this exercise, an instructor could take suggestions from representatives of each group about: (a) content, (b) arrangement, and (c) methods of presentation. Later, this exercise could provide the basis for a discussion on the essential requirements in scientific writing generally – not just in writing instructions.

2   Personal records

Writing helps you to remember

Most working scientists and engineers keep a diary to help them to remember what they must do and what they have done. They make notes in meetings and during investigations; and when they need to be reminded how to use an instrument or apply a technique, they follow detailed instructions, prepared by other people. However, a student's first use of writing as an aid to remembering is when making notes in organised classes and in private study.

Making good notes

The kind of notes taken by students in organised classes will depend on the way information is presented. Some lecturers speak almost as in dictation and the students record nearly every word. In other lectures the students make few notes, but these carefully selected headings and sub-headings, definitions, phrases, words, numbers and abbreviations serve as memoranda. It is from such lectures that they are likely to learn most: they are not fully occupied in writing and have time to think, select and note only the essential points.

A lecturer's task is not to tell students all that they need to know, but to provide a digest of the essentials of the subject supported by examples, to discuss problems, hypotheses and evidence; to explain difficult points, concepts and principles; to refer to sources of further information; and to answer questions. In this way the lecturer acts as a pacemaker for students who, by listening, can move forward more quickly than they could by reading alone. However, students will find it easier to make notes during a lecture if they have done some preliminary reading and have attended and understood the earlier lectures in the same course.

When making notes in an organised class the first thing to write, at the top of a new page, is the date and a title. In deciding what to write next, respond to the way the class is organised. You may decide to record much of what is said if the information is not readily available in a textbook, or if you may be expected to recall details in assessed course work or in examinations. In other classes you may find it most rewarding to listen, to make a few notes, to contribute to discussions and be ready to ask questions.

A good lecturer may start by outlining the lecture. If it has been well prepared, information and ideas will be presented in an appropriate order, and each student's notes will be similar to the topic outline prepared by the lecturer when deciding what to say. For the student they will be a record of the lecturer's main points, and a reminder of the supporting arguments – even if these are not recorded.

Lecturers who provide handouts in every class, which students come to expect and which serve as a substitute for the students' own notes, deprive their students of opportunities to develop the ability to listen carefully, recognise essentials and make their own brief notes. This ability to summarise is an essential skill (see Table 1.1) not only in study, but also in any other employment (for example, in interviews, during telephone conversations and in meetings).

Making notes is an aid to concentration, and students must either make notes as they follow the lecturer's argument and explanation or make fewer notes while they are listening and then go to their books. Whichever method is adopted, they should learn during the lecture and should be ready to ask or to answer questions at the end.

Writing helps you to observe

Observation is the basis of science, and preparing an accurate drawing (see Figure 10.1) or completing a data sheet helps to focus attention on an object or event. Writing, using words, numbers and appropriate units of measurement, is also necessary for precise description – and is an aid to observation and to learning.

In a description, proceed from the general characteristics of an object to the details (as in a definition, see page 66) or perhaps from the outside to the inside. When one point has been covered adequately, look for something else to describe. This will help you to see more and your description will not be confined to the most obvious things. As a trained observer you should try to miss nothing.

The revision of a description provides an opportunity for the rearrangement of observations so that there is a clear distinction between the most conspicuous features and the detail, or so that events are described in chronological or in some other appropriate order, and so that attention is drawn to different observations that seem to be related.

Keeping a record of practical work

During an investigation records should not be made on odd scraps of paper, but in a laboratory or field notebook. Like a diary, this is a permanent record of what is done each day (for example, of the amount of each ingredient in any mixture, the methods used in all preparations, the arrangements for the standardisation of the conditions for each investigation, the instrument numbers, the temperature and atmospheric pressure if these are relevant, and any safety precautions).

The organisation of numerical data should start as they are recorded, on carefully prepared data sheets. These tables should be pages in the notebook, or they should be securely fixed to the appropriate page. The first column of each table (the stub) may have a heading *Date* or *Time* and each of the other columns must have a heading to indicate what was observed or measured each time entries were made, and the units of measurement (see page 93). Preparing a data sheet, when the work is planned, helps you to decide what is to be recorded. During the investigation, a data sheet is an aid to observation. It directs attention to the readings required, helping to ensure that data are recorded in order and at the right time so that a complete record is kept. Then, after the investigation, because the data are neatly arranged, the data sheet facilitates the perusal and analysis of data to derive the results that may be included in your report of the work.

Your notebook is also the place for a drawing of the apparatus, for circuit diagrams, and for line drawings made as records of observations. These should be completed in the class (on unlined paper) and should include a scale bar – as on a map. Notes are normally required to supplement the drawings: the practical class being an opportunity to observe, think, select, record, learn and remember, not an art lesson. Together, a student's drawings and notes should be useful for reference and for revision prior to tests and examinations.

Laboratory and field notes should not be made in rough and copied later. This wastes time and mistakes may be made in copying. It is best to make neat records and to write in carefully constructed sentences. Every note should be dated. The date cannot always be remembered and it may assume great importance later, indicating the order in which things were done. The date is also the key to records made at the same time by other people, of such things as weather, day-length and the state of the tides. For the same reasons, keep a record of the starting time, of the time when each note is made during an observation, and of the time the observation ends.

Details such as these will be required if the work is to be repeated and for the *Materials and Methods* and *Results* sections of a report, and should not be trusted to the memory. It may not be possible to write your report if some detail has not been recorded. Things which seem unimportant during an investigation may prove to be important later. A laboratory notebook should then provide the information about why, how and when, observations were made. A field notebook should provide a similar record not only of why, how and when, but also of where observations were made. If data recorded in a notebook are copied into a computer file, there is a possibility that mistakes will be made in copying. The notebook containing the original data should not, therefore, be discarded: it should be stored in a safe place.

The report on any investigation, whether this is a class exercise completed in one afternoon by a student or several months' work in an industrial or research laboratory, must be based on records prepared during the investigation. It will be easier to write the *Introduction* and *Discussion* sections of the report if the reasons for starting the work, and notes on the development of arguments and hypotheses, are recorded during the investigation.

The use of a bound notebook, with each page dated and with times recorded, helps to ensure that all notes are in chronological order. If loose-leaf notes become disarranged, and each page does not include dates and times, it may not be possible to put them together again in the correct order, or if a notebook is lost during an investigation that has taken weeks, months or years, much time and money has been wasted. Research students and working scientists are therefore advised to use a bound notebook and to keep a photocopy or carbon copy of each page on loose-leaf in another place. Carbon copies have the advantage that they can be made anywhere, as notes are made (at any time). Similarly, back-up copies of computer files should be kept in another place (see page 190). Those who fail to take these basic precautions may lose irreplaceable notes (see Figure 2.1). Such losses, which are never expected, may result, for example, from fire, flood or theft, losing luggage on a journey, or the accidental deletion of computer files.

In work requiring special equipment or facilities do not dismantle the apparatus until your observations have been completed, your data analysed and the first draft of your report written, so that if necessary, you can make further observations at any stage during the work.

In any investigation it is very important to record observations that are not expected, because it is from these and from experiments that go wrong that we may learn most. In his *Life and Letters*, Charles Darwin (1809–82) wrote:

> I had also, during many years followed a golden rule, namely, that whenever a published record, new observation or thought came across me, which was opposed to my general results, to make a memorandum of it without fail and at once; for I had found by experience that such facts and thoughts were far more apt to escape from memory than favourable ones.

Where did I put my notes?

Figure 2.1   Keep a carbon copy of each page of your notebook in a safe place.

It is worth remembering that many discoveries are made as a result of unexpected observations in experiments designed for some other purpose (see Beveridge, 1968). This is because we can make plans only on the basis of existing knowledge – our previous experience. We can recognise problems, speculate, investigate, formulate hypotheses, and test hypotheses by experiment, but we cannot plan for something that could not have been foreseen.

Students are advised to devote their time in laboratory or field work to: (a) thinking about the reason for the investigation; (b) completing the necessary practical work; (c) recording observations (data) and analysing data; and (d) making notes for possible inclusion in the Discussion section of a report. In other words, each practical class should be used as an opportunity to observe, think and learn. So far as possible, the work should be completed in class. Additional notes may be made after the class, but these should be clearly distinguishable from records made during the investigation, as should amendments to lecture notes made in the light of personal observations recorded during practical work.

Students may not be expected to write a report of every laboratory or field exercise for assessment. Some practical classes are organised to provide opportunities to learn a technique or to see things mentioned in lectures. All that may be required, as an appropriate record in the student's notebook, may be a heading, the date, the written instructions or schedule provided for the class, the numerical data recorded during the investigation on a well-organised data sheet, the steps in any calculations, the results, and perhaps a graph, line drawing or annotated diagram. Any original observations, whether these are recorded as numbers, words or lines on a drawing must, however, be recorded in class – when the observations are made.

Writing helps you to think

We think in words, and in writing we try to capture our thoughts. Writing is therefore a creative activity that helps us to sort out our ideas and preserve them for later consideration. The following comments about writing as an aid to thinking were written by people with very different interests.

In *A System of Logic* (1875), John Stuart Mill wrote that 'Hardly any original thoughts in mental or social subjects ever make their way among mankind, or assume their proper importance in the minds even of their inventors, until aptly selected words or phrases have, as it were, nailed them down and held them fast.' H. G. Wells began his novel *The Passionate Friends*, published in 1913, by saying that '. . . the toil of writing and reconsideration may help to clear and fix many things that remain a little uncertain in my thoughts because they have never been fully stated'; and in his lectures *On The Art of Writing* (1916) Quiller-Couch pointed out that 'Words . . . are the only currency in which we can exchange thought even with ourselves.'

For English-speaking people, English is the only means of expression by which they become articulate and intelligible human beings (Sampson, 1925). When, therefore, someone says 'I'm no good at English' what he or she really means is 'I am no good at thinking straight, I can't talk sense, I'm no good at being myself' (Strong, 1951). Clearly, English is not like other school subjects: it is the condition of academic life. So, the teaching of English is the point at which formal education should start, with every teacher a teacher of English.

F. W. Land (1975) in *The Language of Mathematics* expressed the view that a 'command of language is best obtained by using it as a vehicle for disciplining and recording thought and stimulating imaginative thinking'. Writing helps people to arrange their thoughts and to plan their work, and even educated people write badly if they have not thought sufficiently about what their readers need to know or about how best to tell them.

Essay writing as an aid to thinking

The value of writing as an aid to thinking is indicated by the use of the essay as an aid to learning, by the contribution of the written report to project assessment in colleges and universities, and by the place of the thesis in a student's preparation for a higher degree examination. It is still true to say that 'One of the duties of a university is to give instruction, but it is our higher function to teach our students to think, and of this accomplishment the essay or thesis is the chief evidence' (Albutt, 1923).

Writing a progress report as an aid to thinking

In project work and research, as in a class or field exercise for students, writing should be considered as part of the investigation, not as an unwelcome task to be undertaken at the end of the work. Science and writing are not separate and successive tasks; and those who do not start to write until the end of an investigation make their work unnecessarily difficult. It is best to start writing before the investigation, with a clear statement of the problem, the materials and methods to be used, and the hypothesis to be tested in any experiment. An outline prepared at the start should be revised as the work proceeds. This will provide practice in the art of writing and help to draw attention to any flaws in reasoning or to different interpretations of evidence (Beveridge, 1968) and to further work that may be needed. In this way, as information is added under each heading, the draft remains an up-to-date progress report.

A progress report may be requested by a project supervisor, and may be required by those supporting the research, but it is of greatest value to the writer. It helps the investigator to think about the work, to plan additional observations, to avoid irrelevant and therefore time-wasting distractions, to see the project as a whole, and to recognise when the work is complete.

Capturing your thoughts

It is a good idea to carry a pencil and a few sheets of notepaper, or a pocket notebook, so that otherwise fleeting thoughts can be recorded. Make a note of ideas and associations, and of possible further investigations, as they come to mind – so that they may be considered later and not forgotten.

Being well organised

Thinking about, deciding upon and recording your long-, medium- and short-term objectives will help you to organise your life and ensure that in study or in any other employment tasks are completed in order of priority. Making a list of things to be done each day (both at work and for recreation) is the first task in establishing an order of priority in study and in any other employment. Similarly, in planning a composition (see *Thinking and planning*, Chapter 5) or a talk (see *Talking about science*, Chapter 14) listing topics that must be included and numbering them as you decide on the order of presentation will help you to organise your work.

Improve your writing

Use writing, in all the ways considered in this chapter, to help you with your work.

Use wide-lined A4 paper for all your written work

Students are advised to use A4 paper (210 × 297 mm) and to use wide-lined paper for all written work and unlined paper for drawings. Wide-lined paper is provided in most examinations, and is useful for all assessed course work and for personal records. Narrow-lined paper is unsuitable because there is no space between the lines of writing for comments, minor additions and corrections.

Some students use one bound notebook for all their lecture notes, then copy them out into separate files kept at home. Others use a separate notebook for each course of lectures so that they do not have to copy the notes later; but this method is cumbersome. Most students, therefore, prefer to carry one loose-leaf pad. They start each new aspect of their work on a new sheet, leave adequate margins, and space their notes so that there is room for alterations and additions.

With loose-leaf pages there is no need to copy notes because each page can be treated separately and kept in the most appropriate file, the order of the pages can be changed if necessary, and new pages can be added at any time in the most appropriate place.

Date everything you write

Include the date on every note you make in your own records, and on every communication. Unless everything you write is dated, you may find that you are unable to prepare an accurate and comprehensive account of your work because you are not sure when a crucial entry was made in your personal records – or you do not know the date on which a communication to which you must refer was despatched.

Write as part of each investigation

A record of practical work is probably best kept in a laboratory or field notebook, so that all notes are in chronological order. As far as possible try to write your report as part of your investigation (see page 14). Use writing as an aid to observing, thinking, planning and remembering, as well as to communicating your results. Start each report with your name, the date and a concise title that indicates clearly the purpose of the investigation. Use of the commonly accepted headings will help you to present information in an effective order and will help the reader (see page 133).

However, students who are provided with a schedule for a practical class, including a title, an introduction, and enough details of the materials and methods to be used, should not be expected to waste time copying them – and mistakes could be made in copying. Instead, the schedule should be included at the start of any routine report.

Use writing as an aid to thinking and planning

Prepare a job list at the end of each day's work, with items numbered in order of priority. Revise the numbering, if necessary, during the day as some tasks are completed and new tasks need to be added. A job list will help you to use your time effectively, be in control of your work and recreation, and avoid stress.

Use writing and drawing as aids to observation

Much scientific writing is based on clear and accurate description. A useful exercise is to ask a class of students to describe an event, process or object. There will be considerable differences not only in the quality of the writing, but also in the perception of the thing described (Henn, 1960).

Consider how writing helps you with your work

Having considered what scientists and engineers write (see page 6), a class of students may be asked to suggest which of the different kinds of compositions are aids: (a) to remembering; (b) to observing; (c) to thinking and planning (organisation); and (d) to communication (see Table 2.1).

Table 2.1 What scientists and engineers write

Personal records

Writing as an aid to remembering

Notes made in lectures, in meetings and when reading
Notes of addresses and bibliographic details on index cards or computer files
Case histories
Diaries
Personal memoranda

Writing as an aid to observing

Laboratory and field notes
Data sheets
Descriptions

Writing as an aid to thinking and planning

Lists of things to do (job lists)
Topic outlines for written communications and for talks
Progress reports
Notes of ideas as they come to mind

Communications

Writing as an aid to communication

Applications
Routine correspondence (letters, memoranda, e-mail messages)
Essays, dissertations, theses, papers for publication, articles
Technical reports, instructions, procedures, specifications, manuals
Press releases, books, book reviews

---

3 Routine communications

Your value as an employee depends not only upon your knowledge of science or engineering, but also on your ability to communicate information and ideas. You must communicate with the people you work for, and as you take on the responsibilities of leadership you must pass on clear instructions to others.

It is not enough to have good ideas or to do good work; you must also be able to make other people understand what you are doing, why you are doing it, and with what result. It is easy to make a complicated subject seem complicated, but intelligence and effort are needed if information and ideas are to be presented as clearly and simply as possible.

When you correspond with people inside or outside your own organisation, or write reports for administrators, managers and politicians, or try to popularise science (all tasks that require tact, imagination and an understanding of your readers' needs), you must write simply and persuasively.

Your readers may be specialists in different subjects, and will differ in their interests and education. Some may not use English as their first language. So there will be words in your vocabulary that may not be understood by all those to whom you write. You should therefore choose words that you expect all your readers to know and understand, and try to convey each message as clearly, simply and briefly as you can. Do not allow your main points to be lost in meaningless jargon or padding. Try to ensure that your arguments leading to any conclusions or recommendations stand out from the necessary supporting detail, and that your message cannot be misunderstood.

Letters

Business letters, normally on headed notepaper (letterhead), are used when communicating by post or facsimile (fax) with people outside your organisation, and only in exceptional circumstances for internal communications.

Writing a letter is a good test of your ability to communicate effectively. Because it represents you, and usually also your employer, you should take great care over the content, wording and layout of every letter to ensure that it makes a good impression on the recipient. The organisation of most letters can be improved, and their length reduced, if you make a note of the points you intend to emphasise and then number them in an effective order, before writing or dictating your message.

The basic requirements in writing a letter are the same as in any other communication: who requires the information, why is it required, and what exactly do they need to know (see *Helping your readers*, Chapter 8). Having considered what the reader needs to know, you should try to convey this information in an appropriate order, clearly, concisely and courteously – bearing in mind the recipient's likely feelings on reading your words. The tone of each communication should depend on its purpose, but in scientific writing, most communications will be clear, simple, direct, helpful and informative and none will be aggressive or impolite (see Figure 3.1).

Awaiting the favour of your reply,
I remain.

Figure 3.1 All correspondence should be in standard current English (or standard American); not in the outmoded language still sometimes used in commerce (commercial jargon).

Table 3.1 The layout of a business letter

<table>
<tr><td></td><td>Address of sender</td></tr>
<tr><td>Address[a]
of receiver</td><td>Telephone and
fax numbers</td></tr>
<tr><td></td><td>Date of sending[b]</td></tr>
</table>

Salutation,[c]

Subject heading (underlined)

1   Information required (purpose or main point of letter)

2   Supporting details

3   Conclusion and/or action required

Complimentary close,[d]

Signature

Typed name and position of sender.

Enclosures: a list.[e]

Reference line: initials of person signing letter and those of the typist.

Notes
a   The address should be as on the envelope. It should not be punctuated. A formal letter
    should include the position, not the name, of the recipient.
b   The date should be given in full, without punctuation.
c   The salutation in a formal letter should be Dear Sirs, Dear Sir, *or* Dear Madam, *or*
    Dear Sir/Madam, but in a less formal (more personal) letter the name of the recipient
    should be included in the address and in the salutation.
d   The complimentary close in a formal letter is Yours faithfully *or* Yours truly depending on
    national custom, and in a less formal letter it is Yours sincerely. An initial capital letter is
    used for the first word only. The signature should be legible.
e   Any supporting details, if they are more than a few lines, should be on a separate sheet
    which should have a heading and should be dated.

Most letters are written on one sheet of paper. In a few words, whether the letter is sent by post or by fax, you must convey your message and create the right atmosphere between yourself and the person addressed. To keep each letter short and to the point, any supporting details that are needed should be referred to briefly in the body of the letter but sent as an enclosure (see Table 3.1).

The initiator of any correspondence should state the purpose of the letter either by a clear, precise and specific heading, or in the first sentence. The reply and any further correspondence should begin: 'Thank you for your letter of . . . about . . .' From these beginnings both the writer and the recipient know immediately what each communication is about.

It is good practice to give yourself time to check and reconsider letters that are other than routine, perhaps on the day after you wrote them, and to ask someone in the same organisation as yourself to check them – if you need or should seek advice. Then you may decide that some changes are necessary. Apart from such necessary delay, all correspondence should be dealt with promptly as a matter of courtesy and to increase efficiency.

Postcards are useful for concise messages that are not confidential, for example to request details of an item of equipment or an advertised vacancy; or to confirm that a communication is receiving attention and that you will write again as soon as you can provide the information required. The name and address written on the front of a postcard should not be repeated on the back, which should show only the date, your message, your name and address, and your signature. Neither a salutation nor a complimentary close should be included. Because it may be seen by others, as well as the addressee, when a postcard is used to acknowledge receipt of a letter it should not make public the contents and purpose of the letter. It should include only the date of the letter acknowledged and its reference letters/numbers.

Memoranda

Memoranda, normally on memorandum forms, are for internal use only. Otherwise, the basic requirements are as in a letter. Just as a letter represents both you and your organisation, so a memorandum represents both you and the department in which you work. It need not be impersonal, but it should be direct; giving information, suggestions or recommendations, or indicating clearly the information or action required.

A memorandum, like a letter, should have an alphanumeric reference and a concise subject heading (see Table 3.2). Sub-headings should be used if they will ease communication, and the paragraphs should be numbered.

Table 3.2 The layout of a memorandum

NAME OF ORGANISATION

MEMORANDUM

To: ....   From: ....

Dept. ....   Dept. ....

Your ref. ....   Our ref. ....

                              Date: ....

<u>Subject Heading (underlined)</u>

The paragraphs of your message should be numbered.

1

2

Initials of sender (next to and immediately after last sentence)

Distribution: if names are listed here, write *See distribution* after **To:** (above)

        Action ....

        Information ....

Notes
- Because they are used only within an organisation, memoranda begin with the name of the organisation, not the address. For the same reason, the telephone and fax numbers of the organisation are not included.
- Neither a salutation nor a complimentary close is required; and because the sender's name is stated at the beginning it should not be repeated at the end.
- A memorandum is completed not by a signature, but by the sender's initials.
- The words printed in bold in this table indicate an appropriate layout for a printed memorandum form.

Numbers make the writer: (a) think about what is to be said, and (b) arrange the points in an appropriate order. Then they: (c) draw the reader's attention to each point, and (d) help the reader (who may use the same numbers) to compose a reply.

Each memorandum should be composed carefully: it should be as short as possible but as long as necessary. Most memoranda deal with one topic and comprise one paragraph, or just one sentence. To keep each memorandum short and to the point (see Table 3.2), as with a letter, the main points should be stated concisely on one sheet of paper (with any necessary supporting details or further information referred to briefly but sent as an enclosure).

The memoranda used for routine reporting should, if practicable, be report forms. These save time by helping the writer to know what information is required and by whom; and help recipients know where to look for any detail.

Whatever its purpose, consider carefully to whom each memorandum should be addressed. Sending copies unnecessarily to people who do not require them wastes paper, wastes your time, wastes the readers' time, and indicates a lack of judgement.

Electronic mail

With electronic mail (e-mail), as in a memorandum, the names of both the recipient and the sender precede the message, which starts with a subject heading, and there is neither a salutation nor a complimentary close.

All that is necessary in replying to a communication is to insert your message, ensuring that you include: (a) an alphanumeric reference as part of the subject heading, for purposes of filing – and, if necessary, (b) the name of your organisation and your job title.

Communication by e-mail is easy, but to ensure your meaning is unambiguous every message should be in grammatically correct English, with correct spelling and punctuation. It should not be casual or ill considered (see page 1). Every communication, however short, whether sent by post, fax or e-mail, should be prepared in four stages. Always: *think, plan, write*, and then check and, if necessary, *revise* your work (see pages 40–9).

Unless encrypted by a secure server, e-mail messages can be intercepted. So, never send confidential information by e-mail; and never forward (circulate) a message without considering who is entitled to receive the information it contains. Also, never write anything that might embarrass others or cause offence, and bear in mind: (a) that many employers use security products in an attempt to prevent fraud and other misuses of e-mail; and (b) that e-mail messages may be stored for years in an organisation's back-up files. Messages sent by e-mail are neither as private nor as ephemeral as some people may think.

Correspondence should normally be dealt with promptly (see page 19) but because with e-mail it is possible to reply immediately, upon receipt of a message, the temptation to do so may have to be resisted – for several reasons. First, incoming e-mail messages should be placed in order of priority, with other correspondence and other tasks, on your job list (see page 16). Second, even if you acknowledge receipt of an e-mail message immediately, time should be allocated to any necessary thought, consultation or research before you write a considered reply. Special care is needed to ensure that confidential information is not disclosed inadvertently as a result of replying in haste. Third, and in particular, if any message irritates or annoys you, it is essential that you give yourself the time for reflection that you would have had if an immediate response had not been possible.

As much care should be taken in deciding to whom you should send a particular e-mail message as you would take in deciding who should receive copies of a memorandum. Identical messages can be sent to some people (A) for action, with CC (so-called Carbon Copies) to others (B) for information, and with BCC (so-called Blind Carbon Copies) to others (C), so that those listed under A and B do not know about copies sent to those listed under C.

E-mail makes a major contribution to information overload – a major problem for individuals and for businesses. Like any other communication, an e-mail message should be sent only to the person or persons who require the information it contains – not thoughtlessly to everyone in an e-mail group. The receipt of multiple copies of e-mail messages, like the receipt of e-mail that is not relevant to people in your e-mail group, may be an indication of the sender's poor judgement or a result of the poor management of distribution lists.

In responding to an e-mail message it is possible to 'Send' just your reply or to 'Send (with history)' so that your reply is despatched with associated correspondence. However, it is best just to 'Send' because sending with history results in many people receiving copies of large numbers of earlier communications, which they have to read (to check that they are copies of communications received and read previously, and which they have already either deleted or stored in appropriate files). Employees should not have to waste time checking incoming messages, each day, searching for the few that are correctly addressed and deleting all those they should not have been sent.

Another feature of e-mail, contributing to information overload, is that it is easier to append unprocessed blocks of text from other documents, or worse whole documents, than to trouble to extract and summarise relevant parts before sending just the information the receiver needs.

In short, if e-mail is to contribute to efficiency in any organisation, it must be properly managed and then used with consideration and care.

Improve your writing

Put it in writing

Confirm in writing anything agreed on the telephone, or in other conversations. Misunderstandings are likely unless both parties have a written record.

Ensure each communication is well presented

1   Use unlined white paper (A4 size = 210 × 297 mm).
2   Write legibly or use a word processor. Note that business letters and memoranda are usually printed in single spacing, with double spacing between paragraphs, margins of about 2.5 cm at the top, bottom and sides of the sheet, and with the right-hand margins not justified.
3   Leave one space after a comma, semicolon or colon and two after a full stop.
4   Print on one side of each sheet only, and ensure that the layout is acceptable (for example, see Tables 3.1, 3.2 and 3.3).
5   Do not use two sheets of paper if your letter can be rearranged neatly on one.
6   Use a C4 envelope (324 × 229 mm) for A4 paper unfolded; a C5 envelope (229 × 165 mm) for A4 folded once to A5 size; and either a DL envelope (220 × 110 mm) or a C6 envelope (162 × 114 mm) for A4 folded twice.
7   Fold the paper so that it fits neatly into the envelope.

Keep a copy

Every letter or memorandum, whether received by post, fax or e-mail, must fit into the filing system (records) of both the sender and the recipient. This is why it should deal with one subject only and should have: (a) a clear, precise and specific heading; and (b) a unique alphanumeric reference. So if you have to write about more than one subject, to the same person, each subject should be dealt with in a separate communication – even if these are enclosed in the same envelope. And remember to keep a copy: the easiest way to do this, if the communication is handwritten, is to use carbon paper.

E-mail messages that you should not have been sent should be discarded, as should many others once their contents have been noted, but each important message should be stored with related documents (in date order, with the most recent on top) in the appropriate file.

Prepare an application for employment or for promotion

Preparing an application for an appointment, or for promotion, is an exercise that may prove to be both useful and interesting. Completing an application form, or preparing a *curriculum vitae* (résumé) requires care and consideration for the needs of the reader; and, writing a covering letter is a good place to start considering the essentials of clear, concise, courteous and persuasive writing.

On the one hand, a poorly prepared application may cause an employer to decide not to short-list a suitably qualified applicant (see page 1). On the other hand, a well-presented application, with writing that is clear, concise, courteous, confident and convincing, will create an immediately favourable impression.

Most applications are in two parts: a covering letter, and either an application form provided by the employer or an up-to-date *curriculum vitae*. Write a formal covering letter (as in Table 3.3) unless you know the name of the person to whom you should apply, in which case you may choose to write a more personal business letter. When applying for an advertised vacancy, state where you saw the post advertised, quote the alphanumeric reference included in the advertisement, and ask to be considered for this particular position. Say briefly why you are applying: for example, why you consider yourself suitably qualified, what relevant experience and skills you have to offer, and – if appropriate – why you think you would find the work challenging, interesting and rewarding.

Anything particularly relevant to the post must be emphasised in both the covering letter and the *curriculum vitae*. So, a *curriculum vitae* prepared when applying for one post is unlikely to be suitable – without modification – for use when applying for another.

In the *curriculum vitae* start with your full name, date of birth, nationality and address. If your application is speculative (not in response to an advertised vacancy) state the kind of work you are seeking. Then, if you are a student or have recently qualified, give details of your education, work-experience and outside interests, working from the past to the present under each heading (as indicated in Table 3.4). Alternatively, if you already have experience in employment, start with your most recent post (date started, employer, job title and some information as to the work involved). Then give details of your work for any other employers, in the reverse of chronological order. Summarise your other current interests (especially if these are relevant to your application, for example, indicating

Table 3.3 Example of a letter of application for employment

Your address

Date letter is signed

The Personnel Officer   and posted

Name of employer

and full address

Dear Sir, or Dear Madam, or Dear Sir/Madam (as appropriate)

Please consider this application for the post of . . . .

. . . . (Ref. . . . .) advertised in

. . . . on . . . .

I am just completing an honours degree course in . . . .

at . . . . I was deputy head boy at school, and have

worked in a supermarket and in a factory. I have also travelled in . . . .

and . . . . I enjoy working with other

people and should like to make a career in . . . .

I have a particular interest in . . . . My *curriculum vitae*

is enclosed.

I shall be taking my final examinations in . . . .

Otherwise, I could come for interview at any time convenient to you.

Yours faithfully,

any special accomplishments, an ability to work with others, or leadership qualities). Conclude with details of your highest qualifications, using your judgement as to whether any of your earlier education is relevant.

It is best if your *curriculum vitae* can be fitted on to one side of one sheet of A4 paper, with adequate margins (as in Table 3.4). Any essential

Table 3.4 Layout of a *curriculum vitae* or *résumé*

Thomas Jones		Date of birth:	
Home address:		British, Single.	
Telephone no.			
e-mail address:			

Education

Dates . . . .   High School, ...

Date . . . .   examination results

English	B	Science	C
History	C	Mathematics	B
Geography	A	French	A

Date . . . .   examination results

Further mathematics   B   Economics   B   English   A

Date . . . .   University

Studying English, Economics and Mathematics. Reading for honours degree in Economics. Final examinations in June 2011.

Non-academic interests. At school I was deputy head boy and played rugby for the 1st team. At university I play squash for the 2nd team. I enjoy reading, listening to music and going to the theatre. I have a full driving licence.

Work experience

Date . . . .   Vacation work in a supermarket.

Date . . . .   I had a labouring job with . . .

supporting details mentioned in the *curriculum vitae*, or in the covering letter, should be provided on a separate sheet as an enclosure.

Because your *curriculum vitae* is a summary of all the important events and achievements in your life that are likely to be of interest to an employer, every year must be accounted for. If any are not, it may appear that you have something to hide.

Include the correct form of address, the name, position and address of each of your referees (two unless you are asked for more). One referee should be able to speak of your character and interests, and the other of your suitability for the post for which you are applying. So, always choose your referees carefully – for each post for which you apply.

Each of your referees: (a) must have details of a particular post or know the kind of employment you are seeking; and (b) must have agreed to support your application(s). When they agree to act as referees, ask if they would like you to send them details of each post for which you apply. They may be better able to support your applications if you keep them informed, and let them know of any experience or skill you consider particularly relevant to an application.

In your application, as in any other composition, consider your readers. If the post has been advertised, some details will have been included in the advertisement. You may then write for further details and an application form. The further details will tell you more about the post advertised and about the employer.

If there is an application form, return this, instead of your *curriculum vitae*, with your covering letter (see Table 3.3). Keep a copy of your application (the covering letter plus your *curriculum vitae* or completed application form) for reference.

Making the most of yourself in an application is clearly time consuming, but it is worth spending several hours on this task if you are trying to obtain suitable employment – whether this is for a few months' vacation work or for a post in which you may spend the rest of your working life.

4 How scientists should write

Scientific and technical writing should reflect the way scientists and engineers think and work, and should therefore be in accordance with the requirements of the scientific method.

Characteristics of scientific writing

Explanation

Consider first the needs of your readers. Who are they? What do they know already? What more do they require in the way of information, explanation and examples? Always, in scientific writing your purpose is to explain. What is it? What is it for? How does it work? What have you done? Why was it worth doing? How did you do it? What did you find? What do you conclude?

Clarity

The clear thinking that is necessary for the application of the scientific method (in the statement of a problem, in formulating hypotheses, in planning an investigation and in its execution) should be reflected in the clarity of your writing and in your illustrations (see Figure 4.1).

Completeness

The treatment should be comprehensive. Every statement should be complete. Every line of argument should be followed through to a logical conclusion. Your writing should be free from errors of omission but you should show an awareness of the limitations of your knowledge.

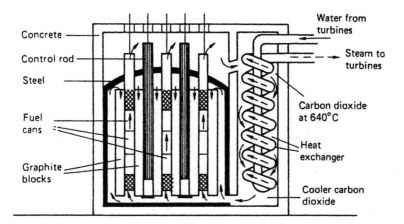

Figure 4.1 The use of an illustration to contribute to clarity and as an aid to explanation: a gas-cooled nuclear reactor represented in a cut-away diagram (simplified and not to scale) to display fuel cells, control rods, flow of carbon dioxide, heat exchanger, flow of water from turbines, and flow of steam to turbines (based on a Hartlepool Power Station leaflet). Diagram from Barrass, R. (1991) *Science*, Basingstoke, Macmillan.

Impartiality

Make clear any assumptions underlying your arguments, for if these are incorrect your conclusions may also be incorrect. Indicate how, when and where your data were obtained, and specify the limitations of your work, the sources of error and probable errors in the data, and the range of validity of the conclusions. Show an awareness of all sides of a question. Try not to be biased by preconceived ideas, and take care not to overestimate the importance of your work. Neither omit evidence that is against your hypothesis, nor undervalue the findings of other scientists when these seem to contradict your own.

Any assumption, extrapolation, or generalisation, should be based on sufficient evidence, and should be in accordance with *all* that is known on the subject. Any assumptions, conjectures, and possibilities discussed, should not be referred to later as if they were facts. Words to watch, because they may introduce an assumption, are: *obviously, surely* and *of course* (see also Table 4.1).

Order

Readers will find your message easier to understand if information and ideas are presented in an appropriate order. The requirement for sufficient

Table 4.1 Phrases used by some people as a substitute for evidence

Introductory phrases	A possible interpretation
It is evident that	I think
It is generally agreed that	Some people think
All thinking people agree that . . .	If you don't agree with me, you must be . . .
It is likely that	I don't have enough evidence to say that
So far as we know	We could be wrong
As you know	This is superfluous
Tentative conclusions	Possibilities

explanation, for clarity and completeness, and for an orderly presentation of information, is most obvious in giving instructions (see Table 4.2).

Accuracy

The scientific method is based on care in planning investigations, care in observation, precision in measurement, care in recording and care in analysing data. Every investigation should be repeatable, and every conclusion should be verifiable. No amount of care in analysing data, or presenting results of the analysis of data, can compensate for lack of care in earlier stages of an investigation or enquiry. Accuracy and clarity in reporting the work also depend on care in the choice and use of words (see Chapters 6 and 7).

Objectivity

Most people respect authority and are reluctant to accept, or even consider, findings or opinions that conflict with existing beliefs. This may be a problem for anyone who has something new to say. In science, statements should be objective (based on evidence), not subjective (based on the imagination or unsupported opinion). So, avoid excessive qualification. Words and phrases that should cause you to think again include *possible, probably, it is likely that,* and *is better referred to as.* Ask yourself: Have I considered the evidence sufficiently? Is there enough evidence for the qualification to be omitted? If not, further observations may be needed before your work can be reported. The latter *possibility seems quite probable.*

When no more information is available on any point, the need for further work may be mentioned. Do not reason from lack of evidence against a hypothesis or state an opinion as a fact. Do not mistake a widely accepted opinion for a fact, or state the opinion of an authority as if it were a fact. Rely on evidence, not authority.

In scientific writing nothing should be implied or left to the reader's imagination. The novelist, journalist or advertiser, to drive home a point, may repeat, exaggerate or understate a case. None of these techniques is available to the scientist or engineer who must tread a more difficult road and convince readers by evidence clearly presented and by logical argument.

Anthropomorphic expressions

Scientists should not endow inanimate objects, or even living organisms other than people, with human attributes. They should not, as students preparing written work for assessment or as working scientists and engineers writing scientific papers for a professional audience, use personal pronouns when referring to animals other than people.

Scientists should not write that *the results suggest*, or that *another possibility suggests itself*, or that *an experiment suggests*, because none of these things can suggest. They should not write that *the data pointed* to the fact (meaning they provided evidence in support of the hypothesis) because data do not point. They should not write *from the point of view of numbers*, because numbers do not have a point of view, or *from the standpoint of soils* (because soils do not stand). In everyday conversation we may say that a car *does not like* a steep gradient, but scientists should not allow such expressions of human emotion to *creep* into their writing.

Teleological expressions

A teleological expression is one in which the end point of an activity or process is stated as if it were a goal. In everyday conversation, for example, we may say the sun is *trying to burst through* the clouds (although we know that it is not). Such misleading teleological arguments are especially common in writing about living organisms. For example, in each of the following sentences a result of evolution (an end point) is stated as its cause (as a goal):

1 Some animals are camouflaged *so that* they cannot be seen by their enemies.
2 A bird has wings *for* flying and a beak *for* feeding.
3 Many insects bury their eggs *in order to* protect them from . . .

A teleological expression may look like an explanation, and so may cause even the writer to think that something has been explained when, in fact, words have been used as a smoke-screen to obscure a lack of understanding. In this way the use of teleological expressions may make less obvious the need for further research and so obstruct the advancement of science.

For a more detailed explanation as to why anthropomorphic and teleological expressions are unacceptable in scientific writing, see Kennedy (1992) *The New Anthropomorphism.*

Simplicity

In choosing between hypotheses, the scientist is asked to prefer the simplest explanation that is in accordance with all the evidence. This basis for choice (that entities must not be unnecessarily multiplied) was suggested by William of Occam, a theologian, in the fourteenth century, and is known as Occam's Razor.

Simplicity in writing (and in illustrations, see Figure 4.2), as in a mathematical proof, is the outward sign of clarity of thought. Scientists should write direct, straightforward prose, free from jargon, verbosity and other distracting elaborations.

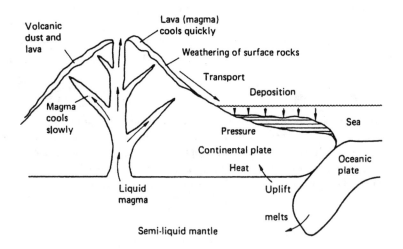

Figure 4.2 The use of an illustration to convey an idea: to help readers visualise something that cannot be seen. Rock formation and rock destruction: continental and oceanic plates of Earth's crust represented on the semi-liquid mantle. A key principle in geology, the principle of uniformity, is that *what is happening now on Earth has happened similarly in the past.* Diagram from Barrass, R. (1991) *Science*, Basingstoke, Macmillan.

Scientific writing

Napley (1975) in *The Technique of Persuasion* advised those advocates who would best serve their clients to present their case in order, with integrity, clarity, simplicity, brevity, interest, and with no trace of pomposity. Explanation, clarity, completeness, impartiality, order, accuracy, objectivity and simplicity are given here as basic requirements in scientific and technical writing. The writing of considerate authors also has the following characteristics:

Appropriateness: to the subject, the reader and the occasion.

Brevity: showing an awareness of all sides of a question; maintaining a sense of proportion.

Consistency: in the use of names, technical terms, abbreviations, numbers, symbols; and in spelling and punctuation.

Control: paying careful attention to arrangement, presentation and timing so that you are always in control – affecting the reader in a chosen way.

Interest: holding the reader's attention.

Persuasiveness: convincing the reader by evidence *forcefully* presented.

Precision: exact definition supported, as appropriate, by counting or by accurate measurement.

Sincerity: the quality of frankness, honesty, humility.

Unity: the quality of wholeness, coherence.

Improve your writing

Criticise other people's writing

Study the following extracts, from work published by scientists and engineers who were, presumably, trying to do their best work. Detecting faults in the work of others should help you to improve your own.

Example 1

The complaint of examiners that students cannot write good English applies, I think, mainly to science students . . . As their abilities lie outside literature, it is not surprising that science students write badly.

SOME FAULTS

1   An opinion is expressed and later stated as a fact.
2   The author gives no evidence in support of the implication that students are good at either literature or science.

Example 2

Under present day conditions there can be little doubt that nitrogen is perhaps the most important factor in feeding the world. It is not necessary to stress the fact that . . .

SOME FAULTS

1   There is excessive qualification in the first sentence.
2   In the second sentence the writer is about to stress something which does not need to be stressed.
3   If something is stated as a fact, it is not necessary to call it a fact.

Example 3

The last ten years have seen changes in teaching of a magnitude unequalled in any previous period of our educational history. Such advances have necessitated a monumental expenditure of money and human resources, and it is interesting to note that whereas in countries like the United States . . .

SOME FAULTS

1   Years cannot see. See *Anthropomorphic expressions*, page 33
2   Of a magnitude unequalled *means* unequalled.
3   In any previous period of our educational history is tautological; it should read *in our educational history*.
4   Changes are later referred to as advances.
5   Advances do not necessitate.
6   Expenditure cannot be monumental.
7   The words *it is interesting to note that* can be omitted without altering the meaning of the sentence.
8   Are any countries, other than the United States, like the United States?
9   The first sentence refers to education in Britain between 1964 and 1974. Is this statement true?

Example 4

Safe and efficient driving is a matter of living up to the psychological laws of locomotion in a spatial field. The driver's field of safe travel and his minimum stopping zone must accord with the objective possibilities; and a ratio greater than unity must be maintained between them. This is the basic principle. High speed, slippery roads, night driving, sharp curves, heavy traffic and the like are dangerous, when they are, because they lower the field zone ratio.

SOME FAULTS

1 The writer's meaning is not clear. Presumably it is that a driver should always be able to stop within the distance that can be seen to be clear?
2 The writer seems to have tried to make a simple subject unnecessarily complex.

Example 5

Much of the Romagna of Italy, for instance, which was fully populated in ancient times, was only restored to its ancient population and productivity by great efforts in the present century.

SOME FAULTS

1	fully populated	How many?
2	in ancient times	When?
3	only restored . . . by	restored . . . only by
4	to its ancient population	Very old people!
5	and productivity	As productive as in ancient times?

Criticise your own writing

Preparing a set of instructions, using words alone or words supported by effective diagrams, drawings, photographs or samples, provides a good introduction to the essentials of scientific and technical writing.

Reconsider your set of instructions (prepared as an exercise, page 7). Are they complete? Are they arranged in order of performance? Are they numbered to emphasise the separate steps? Would all who could be expected to use them understand what to do, and by following your instructions satisfactorily complete the task? Does your set of instructions have all the characteristics of scientific and technical writing considered in this chapter?

Table 4.2 How to write a set of instructions[a]

Stages	Instructions		Essentials
Think			
	1	Consider who may use the instructions, and how they will be used.	Consideration for the reader
	2	Ensure you can complete the task well yourself.	Knowledge and understanding
	3	Precede the instructions with any necessary explanation, words of caution, warning or possible danger.[b]	Safety
	4	Give the instructions a concise but informative heading (as above).	
	5	List any materials or equipment required.	
	6	Break the task into steps: the things to be done, explaining the action required, at each step.	**Explanation**[c]
Plan			
	7	Arrange the steps in order of performance, so that completing the last step completes the task.	**Order**
	8	Include photographs, drawings or diagrams, intended to help the user, next to the instructions they illustrate.	**Appropriateness**
Write			
	9	Write in the imperative (with each step one instruction or command), as in this list. [d]	
	10	Make each instruction as simple as possible.	**Simplicity**
	11	Write each instruction as a complete sentence, using words users will not misunderstand, to ensure it is unambiguous.	**Clarity**
	12	State any safety precautions immediately before any step at which special care is needed, preceded by the word CAUTION, the word DANGER or the word WARNING, as appropriate.[b]	Safety
	13	Number the steps, to draw attention to the action required at each step.	
	14	State any observations, to be made at each step, that indicate a satisfactory outcome.	
	15	Express each quantity mentioned as a number and an SI unit of measurement, unless other units are marked on the equipment to be used.	**Precision**

Table 4.2 Cont.

Check

16 Undertake the task, following your instructions, to check that they are accurate, in order of performance, and complete. **Accuracy Completeness**

17 Revise the instructions, if necessary.

18 Ask someone else, with experience of the task to undertake the task, following the instructions, and to suggest any improvements. **Coherence**

19 Revise the instructions, if necessary.

20 Ask at least one other person, with appropriate experience, but who has not previously performed the task, to undertake the task following your instructions, and to suggest any improvements.

21 Revise the instructions if necessary.

22 End instructions for use within an organisation, if appropriate, with a statement indicating to whom users should send comments or suggestions. For example: 'Let me know if you encounter any difficulties or have any suggestions for improving either the procedure or these instructions.'

23 Sign and date the instructions. In doing this you take responsibility for them. As with any other communication, you should not sign unless you have authority to do so.

Notes

a Anyone writing safety precautions, user guides, operating instructions, technical manuals, protocols, procedures or similar documents, containing instructions, must ensure that they conform to relevant Standards. They must also satisfy themselves that their responsibilities to users under product liability and health and safety legislation have been met. For example, see DTI (1988) *Instructions for consumer products* and the ISO/IEC Guide 37:1995 *Instructions for use of products of consumer interest.*

b The word *caution* draws attention to a low risk (of damage to the product, process or surroundings), the word *warning* to a medium risk, and the word *danger* to a high risk (of injury or death).

c Words printed in bold in this table are also used as sub-headings in this chapter, as essential characteristics of scientific writing.

d Most instructions begin with the verb, indicating immediately the action required; but if an observation or decision has to be made, *before the action is taken*, the sentence may start with 'If . . .' or 'When . . .' (for examples, see Table 5.2).

From the advice included in Table 4.2, can you revise any instructions you have prepared so as to improve the performance of users? Could you suggest ways in which other written procedures produced by your employer could be improved?

5   Think – Plan – Write – Revise

If you would like to improve your writing, you have taken the first step by being prepared to consider how it could be improved. To ensure further improvement and so provide encouragement, you should prepare every composition – however small – in four stages: always *think, plan, write,* and then check your work – and *revise* it if necessary.

Thinking and planning

Before starting any work consider when you must finish. Usually you will be working to a deadline imposed by someone else – a tutor, a supervisor, an examiner, an employer. However, even when working only for yourself, you must consider the time available for the composition in relation to your other commitments. Effective organisation involves working to a self-imposed timetable so that you can meet deadlines imposed by others or by yourself.

The first stages, thinking and planning, will help you to get started and take you well on the way to completing your composition. You will also find your work easier if you do not expect your first draft to be perfect and are prepared to spend time on its revision.

Collecting information and ideas

A common fault in writing is to spend too much time on the early part of a composition. As a result, some ideas receive too much attention simply because they are considered first and others receive too little attention because they come later.

Planning may take a few minutes (as when answering a question in an examination or considering your reply to a routine communication) or much more time may be spent searching for information, discussing your ideas, and organising your thoughts. A good title should help you to define the purpose and scope of the composition, and should inform the reader. Similarly, for a commissioned report, the terms of reference should make clear precisely what is needed, by whom and when. Consider your readers and anticipate their questions. They want relevant information – well organised, clearly presented, and with sufficient explanation. In conversation they would ask one-word questions that you can ask yourself. Rudyard Kipling lists their names in *The Serving Men*. They are:

> What? Why? and When?
> And How? Where? and Who?

Your answers to these questions can never be just 'yes' or 'no'. Record your answers as relevant points come to mind. They will lead to further questions; and you will find that you know much more about many subjects than you at first supposed.

In a few moments of thought and reflection you can usually make a succession of relevant notes. You may use some as topics for separate paragraphs in your composition, with others as supporting ideas, facts and examples in each paragraph. You may leave out other points in your final selection of material, either because they provide unnecessary detail or because you select better examples.

Preparing a topic outline

As you consider the purpose and scope of your composition, and assemble information and ideas, it is a good idea to spread key words, phrases and sentences over a sheet of paper or over the whole of a computer screen (or to write them on separate index cards). Use your main points as headings, and note supporting details and examples below each heading. Then number the headings (or place the index cards in order) as you decide:

> What is the purpose and scope of this composition?
> How is the subject to be introduced?
> What is the topic for each of the other paragraphs?
> What information and ideas must be included in each paragraph?
> Are any tables or diagrams needed? If so, where should they be placed?
> What can be left out?
> What needs most emphasis?
> How can the paragraphs be arranged in an effective sequence?
> How should the composition be concluded?
> Would sub-headings help the reader?

Teachers of English in schools may instruct their pupils not to use headings in imaginative writing; and headings are not used in novels, short stories or other literary essays. But, whatever you are writing, headings will help you to organise your work. So they should be included in your topic outline.

In a composition, headings provide signposts for the reader. Students should include appropriate headings in their answers to questions, in both course work and examinations: (a) to draw attention to the relevance of each part of each answer to the question set; and so (b) making it easy for assessors to see where relevant points have been made and marks should be awarded.

In scientific and technical writing appropriate headings must be used (see also page 133). These headings should be used in preparing a topic outline, to ensure that all paragraphs are relevant to the preceding heading as well as to the composition as a whole.

After collecting all the information you require, if there is time it is a good idea to put your topic outline on one side for a while. Some of your second thoughts may be better than your first thoughts, and you may save time in the end because – even when using a word processor – it is easier to revise a topic outline than to reorganise and rewrite a poorly organised or badly expressed first draft.

Putting your paragraphs in order

After the title and the introductory paragraph, further paragraphs should be arranged so that they lead smoothly to the closing paragraph. An appropriate order may, for example, be chronological or geographical. In a short work it may be an order of increasing importance, or in a long work an order of decreasing importance, or the order may be dictated by the requirements of an assessor, supervisor, employer or customer, or by an organisation's house rules.

The first paragraph is your readers' first taste of what is to come. Here you must capture their interest. Your first paragraph must leave no doubt as to the purpose and scope of the composition (if this is not clear from the title), but there are many ways of beginning (see *How to begin*, page 81)

There should be one paragraph for each aspect of the subject (for each topic). Each paragraph should have a well-defined purpose and be clearly relevant. It should be well ordered. The topic is usually stated (or is apparent) in the first sentence, but in an explanation or argument it may come last. All sentences must provide information or ideas relevant to the topic – but nothing irrelevant – and the first and last sentence should help to link paragraphs so that readers appreciate immediately how your thoughts about one topic lead on to those to be expressed in the next.

Table 5.1 Introductory phrases and connecting phrases that should be deleted from most sentences

---

In considering . . ., it is appreciated that . . .
It is interesting to note that . . . of course . . .
In order to keep the problem in perspective we should like to emphasise that . . .
In conclusion, in relation to . . ., it was found that . . .
It is known from an actual investigation that . . .
The evidence presented in this report supports the view that . . . in the field of . . .
It is not necessary to stress the fact that . . .
It could be said that . . . It goes without saying that . . .
So far we have been discussing what we call . . .
At this point in our discussion . . . we wish simply to emphasise that . . .

Because first and last words attract most attention, never begin or end a paragraph with unimportant words. In particular, omit superfluous introductory phrases (for example: *First let us consider . . . Secondly it must be said that . . . Next it must be noted that . . . An interesting example that should be noted in this context is . . . In conclusion . . .*). Such phrases will be in your mind as you prepare your topic outline, and the numbers in your outline help you to put your thoughts in an effective order, but remember that your outline is for you (as an aid to thinking and planning). It is not for your readers. They require only the results of your thought.

Superfluous introductory and connecting phrases (see Table 5.1) distract attention; they are a fog that obscures your message. The change of topic is clearly sign-posted by the break between paragraphs and in scientific and technical writing the new topic is usually introduced directly and forcefully in the first few words of each paragraph.

Each sentence in a paragraph should convey one thought, and punctuation marks should be used when they are needed to clarify meaning or to make for easy reading. Each sentence should be obviously related to the preceding sentence and to the next. No new statement should be introduced abruptly and without warning. Within a paragraph, therefore, the sentences should be in an effective order so that they hold together and convey your meaning precisely.

Balance is important in writing as in most things. The sentences in a paragraph and the paragraphs in an essay, or similar short composition, like the handle and the blade of a knife, must be balanced in themselves and in relation to one another. Your composition as a whole must be well balanced: ideas of comparable importance must be given similar emphasis.

Paragraphing breaks up the page of writing, provides a pause at appropriate points in the narrative, and helps the reader to know that one thing has been said and that it is time to think of the next. Short paragraphs are also the easiest to read and so they make for efficient communication. However, paragraphs are units of thought, each with one thought or with several closely connected thoughts. So they vary in length.

The topics covered in a short composition, for example in an essay, letter or short report, may lead to some conclusion and/or to recommendations, or may provide the basis for speculation, or may emphasise some aspect of the subject that serves as a link between paragraphs and leads to some hypothesis or theory. However you bring your composition to a close, the end should be obvious to the reader (see Figure 5.1). It should not be necessary to begin the closing paragraph, as do many inexperienced writers, with the words: *In conclusion . . .*

Writing

With your plan complete, the theme chosen and the end in sight, try to write the whole of any short composition at one sitting, using the words that first come to mind. Stopping for conversation, or to revise sentences

The end should be strong, forceful,
convincing and final.

Figure 5.1 The end of any composition should be a rounding off. It should leave the reader with a lasting impression of the work.

already written, or to check the spelling of a word, or to search for a better word, may interrupt the flow of ideas and so destroy the spontaneity that gives freshness, interest and unity to writing. The time to check your work, to consider whether or not it could be improved, is when the first draft is complete.

With your topic outline before you as a guide, you can write with the whole composition in mind – with your message as the thread running through the whole: each word contributing to the sentence, each sentence to the paragraph, and each paragraph to the composition.

Knowing how you will introduce the subject, the order of paragraphs, and how you will end, you can: (a) begin well; (b) avoid repetition by dealing with each topic fully in one paragraph; (c) ensure relevance; (d) emphasise your main points; (e) make proper connections to help readers follow your train of thought; (f) write quickly, maintaining the momentum that makes a composition hold together; and (g) arrive at an effective conclusion.

In short, your topic outline contributes to order and to the organisation that is essential in writing. Only by working to a topic outline – your plan – can you maintain control, so that you present your subject simply, forcefully and with an economy of expression.

In scientific and technical writing, information and ideas should be presented in an interesting and objective way. Avoid figurative language because it may confuse some readers. Use enough words to make your meaning clear – too few will provide insufficient explanation and too many may obscure meaning and waste your readers' time.

The use of the imagination may be encouraged in an English essay, and there is a place for the imagination and for speculation in science (for example, in the formulation of hypotheses) but non-fiction should be impartial, accurate and objective. When the interpretation and assessment of evidence call for the expression of an opinion, this must be clearly stated as such.

Arguments in favour of any conclusion should be based on evidence summarised in your composition. Enough explanation, and where necessary examples, should be included to enable readers to judge their validity. And criticism of other people's work must be reasoned and not based on preconceived ideas for which you can provide no supporting evidence.

To copy the work of a fellow student or of a colleague and present it as your own would be cheating, and if detected would probably be severely punished. Similarly, taking sentences from the published work of others and presenting them as your own is plagiarism (stealing thoughts), and this too is unacceptable. Instead, sources of information and ideas should be cited in the text of any composition, and complete bibliographic details of each publication listed after the heading *References* at the end. By citing sources you acknowledge the work of others, indicate your awareness of its relevance to your own, and inform your readers (see pages 129–30).

Revising

Two processes are involved in written communication. The first, in your mind, is the selection of words to express your thoughts. The second, in the mind of the reader, is the conversion of the written words into thoughts. The essential difficulty is in trying to ensure that the thoughts created in the mind of the reader are the same thoughts that were in your mind. Too often the reader, looking at an ambiguous sentence or at a sentence that is obviously incorrect, must try to work out what the writer meant (see pages 35–7). If you take pride in your work, you must revise carefully. Try to ensure that your words do record your thoughts. Try to ensure that the reader takes this same meaning.

A common fault in writing is the inclusion in one place of things that would be better in another. Indeed, one of the most difficult tasks is to get everything into the most effective order. One reason for this, even after careful planning, is that we think of things as we write – and include them in one paragraph although they would be better placed in another, or even under a different heading.

In writing we use words as they come to mind, but our first thoughts are not necessarily the best and they may not be arranged in the most effective order. Wrong words and words out of place lead to ambiguity and distract the reader's attention, and so have less impact than would the right words in the right place. By further thought, intelligent people should be able to improve their first draft.

Revise carefully so that readers do not have to waste time on an uncorrected first draft which may reflect neither your intentions nor your ability. Read the whole composition aloud to ensure that it sounds well, and that you have not written words or clumsy expressions that you would not use in speech. (See Table 5.2, page 49 and the check lists on pages 151 and 155.)

Thinking, planning, writing and revising are not separate processes, because writing is an aid to thinking. The time taken in planning, writing and revising is time for thought. It is time well spent, for when the work is complete your understanding of the subject will have been improved.

To admit that you need to plan your work, that your first draft is not perfect, that you need to revise your first draft, and that you can benefit from the comments of a colleague or from the advice of an editor, is not to say that you are unintelligent. Even after several revisions you may not appreciate all the difficulties of the reader. It is a good idea, therefore, to ask at least two other people to read your corrected draft of any important communication. Preferably one reader should be an expert on the subject and the other should not be. Coming fresh to the work they will see things that are not sufficiently explained, that are irrelevant (not necessary or out of place) and ambiguous or do not convey the thought that you intended.

Because the quality of your writing reflects upon your employer as well as upon yourself, some employers have a procedure for editing and revising manuscripts. Your employer may also wish to ensure that nothing confidential or classified as secret is reported. You should also remember that in talking or writing about your work you may invalidate a later patent application. If you need advice, consult a patent agent.

The function of a critic is to improve your writing and any comments should be welcomed and should be considered when you revise your work. Because of this, do not ask people to read a draft unless you respect their judgement and can rely on them to give an honest opinion. You are fortunate if you know someone who will criticise your writing as well as the subject matter.

The readers will see mistakes, ambiguities, badly presented arguments, and superfluous words and sentences, which are immediately obvious when someone points them out. They will see good points which require more emphasis (see page 82). Experienced writers learn to improve their first drafts but they can still benefit from a reader's frank comments.

When a paper has been revised, it is a good idea to put it on one side for a while. One way to do this is to send a copy to another reader. There is bound to be a delay and this is the time to get on with something else before you reconsider the composition. The value of a third reader's comments is that they may reinforce the views of the first two readers, and new comments may be made.

An important article or report will probably be typed several times. Every time it is put on one side and then reconsidered, and every time it is read by someone else, further improvements will be made. Each draft should be easier to read, easier to understand, and therefore more interesting than earlier drafts.

To see how hard writing is, even for experienced writers, we have only to study their manuscripts; they are full of alterations, crossings out, additions, loops, arrows . . . And what is hard for them is also hard for us (Vallins, 1964). The apparent spontaneity of easy-reading prose is the result of hard work; for all writers need to correct and improve their first drafts.

Charles Darwin, for example, took great pains with his writing and would laugh or grumble at himself over the difficulty he found in writing English, saying, for instance, that if a bad arrangement of a sentence were possible he would be sure to adopt it:

> When a sentence got hopelessly involved, he would ask himself 'now what do you want to say?' and his answer written down, would often disentangle the confusion.
>
> Francis Darwin (1925), *The Life of Charles Darwin*

Ernest Hemingway wrote the last page of *Farewell to Arms* thirty-nine times before he was satisfied with it. Aldous Huxley said: 'All my thoughts are second thoughts.' H. G. Wells wrote a first draft 'full of gaps' and then made changes between the lines and in the margin. He revised the whole work as many as seven times (Tichy and Fourdrinier, 1988).

Those who write best probably spend the most time criticising and revising their prose; making it clear and concise but not stultified; and ensuring a smooth flow of ideas. However, writing is only one part of a scientist's work and there comes a time when the task of revision has to stop. Furthermore, revision must not be taken so far that the natural flow of words is lost, for language that is artificial in its bluntness and simplicity may lack interest and style. Alan Sillitoe said of *Saturday Night and Sunday Morning*: 'It had been turned down by several publishers but I had written it eight times, polished it, and could only spoil it by touching it again.'

The pleasure to be derived from writing comes from the effort of creative activity – which may lead you to a deeper understanding of your subject. Each composition is original: it is a vehicle of self-expression, a presentation of information and ideas in a way that is peculiar to the writer. No two people will select the same material for inclusion, arrange the arguments in the same way, make the same criticisms, or reach the same conclusions.

Pleasure comes from writing something which will affect other people. The reader may be persuaded or convinced by argument logically presented, or may be annoyed or misled by poor writing. Each communication is a challenge to the writer to present information and ideas directly and forcefully, to help the reader along, and to affect the reader in a chosen way; for this is the purpose of all exposition.

Improve your writing

Of the following suggestions, the first two may be used by teachers of scientific and technical writing as a basis for class work and class discussions, and the next five could be used by teachers of scientific or technical subjects as ideas for set work or for class discussions.

Table 5.2 How to write: four stages in composition

Think

1 Consider the title or your terms of reference.
2 Define the purpose and scope of your composition, if these are not clearly stated in the title.
3 Decide what your readers need to know.
4 If possible, identify your readers and prepare a distribution list.
5 Consider the time available and allocate this to thinking, planning, writing and revising.
6 Make notes of relevant information and ideas.

Plan

7 Prepare a topic outline.
8 Underline the points you will emphasise.
9 Decide on an effective beginning.
10 Number the topics in an appropriate order.
11 Decide how to end.
12 Decide what help you will need with the preparation of diagrams and photographs, editing, copying and binding, or other tasks, and liaise with the people concerned.

Write

13 If your first draft is hand-written, use wide-lined A4 paper with a 25 mm margin. Write one paragraph on each sheet and write on one side only, so that – as with a word processor – you can revise paragraphs or change their order easily.
14 If possible, put other tasks on one side and write where you will be free from interruption.
15 Use your topic outline as a guide.
16 Use effective headings, and keep to the point.
17 Start writing and try to complete your first draft, or one section of a long document, at one sitting, using the first words that come to mind.

Check

18 Does your first draft read well; is it well balanced?
19 Are the main points sufficiently emphasised?
20 Is anything essential missing?
21 Is the meaning of each sentence clear and correct?
22 Does the writing match the needs of your readers, in vocabulary, sentence length and style?
23 If necessary, revise your composition. Then put it on one side for a while to give yourself time for reflection.
24 Read it again to see if you are still satisfied that it is the best you can do in the time available.

Practise essay writing

Instruction on essay writing in schools is likely to be given mainly by teachers of English, who may approve of quite a different style of writing from that required in answering essay-type questions in other subjects – or in most careers.

In an English essay the approach does not have to be systematic; the theme may be developed without formal argument, and imaginative writing is encouraged. The writer may not always strive for clarity: more may be implied than is stated.

An essay in science or engineering is a vehicle for conveying information and ideas: it is a short written account of a well-defined subject. It is clear and decisive, systematic and comprehensive, with the parts signposted by carefully chosen headings (see page 42).

Practice in essay writing will help you to develop your ability to organise your thoughts and present them in English that will be understood by the reader you have in mind. Every teacher of science or engineering should therefore set exercises in writing and should be able to give encouragement and feedback in the form of constructive criticism. Students will find that they learn about their subject at each stage in their writing: from gathering information and ideas; from selecting and arranging their material; from writing; from revising (as if they were correcting and marking their own work); and, if necessary, from rewriting. They should also learn from the markers' comments on their work.

However, an essay is not only an exercise in thinking and writing for students, but also a vehicle in which any writer's thoughts are assembled and organised and conveyed to the reader in a clear, concise and interesting way (for example, a magazine article or review). Whether you are a student or a working scientist or engineer, practising essay writing or writing for publication, either choose a subject that interests you and you know something about or fill gaps in your knowledge and understanding before you start to write, so that you can select, arrange and maintain control of your material.

Study the technique of successful essayists

Consider, for example, the purpose and scope of a leading article in a good newspaper, or an article in a magazine that interests you:

> How did the title capture your interest?
> Does the opening sentence make you want to read the article?
> Reconstruct the writer's topic outline by picking out the topic for each paragraph.

Is each topic relevant to the title?
Are the paragraphs in an effective order?
Do they lead smoothly to an effective conclusion?

Study one paragraph: Can you understand the idea presented in each of the sentences of the paragraph? Are they all relevant to the topic? Why are the ideas presented in this order? Could they have been arranged more forcefully, or more persuasively, in another sequence?

Prepare a topic outline

Write a topic outline for an essay or for a magazine article on a scientific or technical subject that is of particular interest to you. Your plan should comprise key words and phrases, with brief notes on each topic to remind you of information and ideas to be included in each paragraph. Use one sheet of paper or the whole of a computer screen for your rough work, then rearrange your first thoughts as a numbered list to indicate the order of paragraphs.

Academics can help their students by asking for a topic outline to be submitted with each composition handed in for assessment, and then providing feedback on the topic outline as well as on the assessed work.

Consider how examination papers are set and marked

Candidates for examinations should consider how examination questions are set and how they will be marked so that in planning their answers they can attempt to score marks. Each part of an answer, and each paragraph, should be seen as an opportunity to gain marks by adding relevant information and ideas, and by showing understanding.

1   In most examinations marks are divided equally between the questions to be answered so that there are 25 marks per question when four questions are to be answered, and 20 marks per question when five are to be answered. You must, therefore, answer the right number of questions.
2   To be fair to all candidates, examiners allocate the marks which may be obtained for each question according to a marking scheme. This scheme is a topic outline:

If the question is set in parts, a certain number of marks will be allocated to each part of the answer.
Whether or not the question is set in parts, the examiner will expect the students to refer to, and to show their understanding of, all those things which are relevant to the answer.

A student who does not answer all parts of any question or who gives an answer which is otherwise incomplete, cannot score full marks on that question.

In examinations make things easy for the assessors

Answering questions in an examination

1   Do not make vague statements. Give reasons and examples.
2   Do not leave things out because you think they are too simple or too obvious. Do not include anything that is irrelevant, but make sure that everything relevant is included and clearly explained, however briefly, to show your knowledge and understanding. Remember that the examiner cannot assume that you know things, and can give you marks only for what you write.
3   If you include anything that is not obviously relevant, explain why it is relevant. An examination is not simply a test of your ability to recall facts and ideas. It also provides an opportunity to show your ability to distinguish relevant from irrelevant material.
4   Make things as easy as you can for the examiner. Write clearly. Get to the point quickly and keep to the point. Plan your answer so that it is well organised and well balanced, and so that, without digression or repetition, you can say everything you wish to say in the time available.
5   Make sure that any diagram is simple so that it can be completed quickly and neatly. Use coloured lines to represent different things. Do not waste time on shading.
6   If you have read a book, a review or an original paper, relevant to your answer, refer to this by giving the name of the author (and the date of publication), to show the source of your information (see pages 129–30).
7   If a question is set in several parts, answer all the parts and answer them in the order in which they are set – because the examiner is expecting to mark the parts in this order.
8   If a question is set in several parts, answer each part separately; and if the parts are numbered use the same numbers to indicate the separate parts of your answer. If the parts are not numbered, use appropriate headings to draw attention to the parts of your answer.
9   If you are asked to discuss, you must discuss all sides of the question and refer to any unresolved problems.
10 If you are asked to compare, you should also refer to any differences (even if the question does not say compare and contrast).

11 Write in black or blue-black ink. Do not use coloured inks that could be confused with the examiner's comments and corrections. To correct any mistakes, draw an oblique line through any letter or a horizontal line through any word that you delete; then write your correction immediately above the deletion, in the space between the lines of writing. Do not waste time erasing mistakes with a rubber or painting over them with correcting fluid and waiting for it to dry.

Make good use of your time in examinations

1 Read the instructions at the head of the question paper.

2 Read all the questions carefully until you are sure that you know what is required in each answer. Then select the questions which you can answer most fully. Otherwise you may realise, after leaving the examination, that you could have answered another selection of questions and obtained better marks.

3 Before you plan your answer read the question again to make sure that you understand what is required. Answer the question you have been asked and not a similar question which you were hoping for.

4 For essay-type questions, plan all your answers quickly at the beginning so that you can reconsider each topic outline before you start to write.

5 Answer the required number of questions. If all questions carry the same number of marks, divide your time equally between them. Do not spend more time on those questions you know most about. Remember that it is easier to score half marks on a question that you do not know much about than it is to score full marks when you think you can write a good answer. The first few marks are the easiest to obtain, with a little thought, if you know anything about the subject. But a little extra time spent on a question, upon which you have already spent long enough, is likely to be less rewarding.

6 Keep an eye on the time. Allow a proportion of the time available for reading all the questions at the beginning, for planning your answers, and for reading through your work at the end to correct any slips of the pen and to add important points that you did not remember the first time through.

7 Do not waste time. Arrive at the examination before the start. Do not waste time during the examination. Do not leave before the end.

6   Choosing words

Word games, for example crossword puzzles, are popular because there is fun to be had from words; and the habit of consulting a good dictionary whenever you come across a word that you do not understand can be a life-long source of enlightenment and pleasure.

Our interest and pleasure in words is not surprising, for when we speak or write we are putting our thoughts into words. Indeed, the use of words is even more fundamental; for without words, we cannot think. We are limited in our ability to think by the number of words at our command. If we have a large vocabulary and can construct effective sentences and paragraphs, we are better able to express ourselves.

We write so that we can tell others what we think, but if we use words incorrectly – or use words that our readers do not understand – we shall be misunderstood. Clearly, we must think about words so that we can use them correctly and choose those that we expect our readers to know.

English is the language used by most scientists and engineers for international communication; but whether people are using English as their first or second language, they are most likely to understand plain words in simply constructed sentences. So if you wish to be widely understood, write in standard English (see page 186), try to express your thoughts as clearly and simply as you can, and try to be consistent in your spelling and in your use of abbreviations, capital letters and hyphens throughout any document.

The meaning of words

One of the delights of English is its rich vocabulary, which provides a selection of words with which to express our thoughts. This is why, at school, pupils are encouraged to increase their vocabulary, starting with short words they find easiest to say and write. But students, and working scientists and engineers, having acquired a large vocabulary, should choose words that convey their meaning, and should try to match their writing to the needs of their readers – each with a different and many with a smaller vocabulary.

No two words are identical in meaning, and the use of one when another would make more sense will not help your readers. When *The Times* newspaper reported that Rudyard Kipling was to be paid £1 a word for a story, a student at Oxford University sent him £1 and asked, 'Please send one of your best words'. Kipling replied, 'Thanks'.

The right word may not always come so easily to mind, and people who do not know what to say or have too few words at their command may choose the wrong word, or a word that may sound good without adding to the sense (for example: deprived, dialogue, escalation, hopefully, integrated, meaningful, obscene, overall, paradigm, relevant, traumatic), or a hackneyed phrase (for example: *pushing back the frontiers of knowledge, last but not least, at the end of the day, in the last analysis,* and *all things being equal, we hope to see the light at the end of the tunnel*). Instead of such over-used words and phrases (and idiomatic expressions, see Table 7.2) always take the trouble to use words of your own choosing to convey your own thoughts.

The habit of writing a word in quotation marks (see 'out', page 57) to indicate that it is not quite the right word, or that it is not used in the commonly accepted sense, or that more is implied than is said, is likely to confuse some readers and so is to be avoided in scientific and technical writing. Instead, choose the word or words that convey your meaning precisely and, if in doubt, refer to a dictionary to make sure you are using the right word.

The meanings of words may change so much that they lose their value, or what was an incorrect use may come to be accepted. However, scientists should not lead the way in giving a new meaning to a word, unless it is to be used in a new context and there is no possibility of confusing readers.

Some words commonly confused

To illustrate the need for care, here are some words that many writers confuse, and so misuse. Concise comments are included to make clear differences in meaning.

    Accept (receive) and *except* (not including).
    Advice (suggestions) and *advise* (to give advice).
    Affect (to alter or influence) and *effect* (to bring about, or a result).
    Alternate (to perform by turns), *alternately* (first one thing then an alternative, repeatedly, as with a light flashing on and off), and *alternatively*

(referring to one thing as an alternative to another). Strictly, therefore, one thing may be an alternative to another, but with more than two to choose from you have a *choice*, not an alternative.

Amount (mass or volume of something measured) and *number* (of things counted).

Complement (to add to or make complete) and *compliment* (to congratulate, or an expression of regard).

Complementary (adding to) and *complimentary* (without charge).

Continuous (non-stop) and *continual* (repeatedly).

Council (a committee) and *counsel* (advice, an advisor, or to advise).

Data are facts of any kind, which may be measurements recorded as numbers (numerical data) or other observations recorded as words, whereas *results* are obtained from data by deduction, calculation or processing. It is incorrect, therefore, to speak of raw data, but correct to refer to original observations as original data.

Dependant (one who is dependent on another) and *dependent* (relying on).

Discreet (prudent, wary) and *discrete* (separate, distinct).

Disinterested (impartial) and *uninterested* (not interested).

Farther (more distant) and *further* (additional).

Fewer (a smaller number of) and *less* (a smaller mass of): for example, it is possible to have fewer people, but not to have less people.

Imply and *infer*: a speaker or writer may imply (hint at) more than is actually said or written, and from this the listener or reader may infer (guess or understand) the intended meaning.

Its (possessive), indicating that it belongs to someone or something, and *it's* (colloquial) a contraction, meaning either *it is* or *it has*.

Licence (permission, leave, liberty, a permit) and *license* (to authorise).

Majority (the greater number; the excess of one number over another) and *most* (nearly all). In an election a majority is the number by which the votes for the winning candidate exceeds those for the candidate who comes second. If you read that 'the majority of writers use word processors', does this mean nearly all writers use them? Does anyone know what proportion of writers use them? Would it be better to say simply that many writers use them? What is the difference, quantitatively, between the majority and the vast majority? Clearly, the word majority is used by some writers – when they are unable to be precise – as a substitute for evidence (see *Using numbers as an aid to precision*, page 91).

Method (how to perform a task) and *methodology* (the study of method).

Oral (spoken) and *verbal* (using words). In speaking face to face we use facial expressions and other body language (non-verbal communication) as well as words, whereas in writing we must rely on words alone.

Parameter (a characteristic of a population, estimates of which are called statistics) and *perimeter* (a boundary).

Practicable (something that could be done) and *practical* (not theoretical). A project may be considered impracticable because it is not cost effective, but to say that something is not a practical proposition means that it could not be done.

Practice (a customary action, a performance, a business) and *practise* (to exercise, to perform).

Principal (first in rank, main, original or capital sum) and *principle* (a fundamental truth, a law of science, or a rule of conduct one is unlikely to break – as in a matter of principle).

Results should not be confused with data (see *Data*, page 56).

Since (from that time) and *because* (for this reason).

Stationary (not moving) and *stationery* (writing paper).

Their (indicates possession, as in their office, in their own time, their suffering) and *there* (used with the verb *to be*, as in: *there is, there are, there was, there were*; also used to mean *in that place* – as in *over there*).

While (at the same time as) and *whilst* (although).

Who's (colloquial, meaning who is) and *whose* (possessive).

Within (enclosed by) and *in* (inside). Many people use the word *within* when the word *in* would serve their purpose better: for such people, apparently, the word *in* is 'out'. Something may be within these walls or within the bounds of possibility, but unless some such limits are intended the word *in* should be preferred.

Your (possessive) and *you're* (colloquial, meaning you are).

Other words commonly confused are: *admitted* (for *said*), *anticipate* (for *expect*), *always* (for *everywhere*), *centre* (for *middle*), *centred around* (for *centred on*), *circle* (for *disc*), *it comprises of* (for *it comprises*, or *it consists of*), *degree* (for *extent*), *either* (for *each* or *both*), *except* (for *unless*), *fortuitous* (for *fortunate*), *generally* (for *usually*), *homogenous* (for *homogeneous*), *if* (for *although*), *importantly* (for *important*), *improvement* (for *alteration* or *change*), *informed* (for *influenced*), *lengthy* (for *long*), *less* (for *fewer*), *limited* (for *few, small, slight* or *narrow*), *myself* (for *me*), *minor* (for *little*), *natural* (for *normal*), *optimistic* (for *hopeful*), *optimum* (for *highest*), *percentage* (for *some*), *quite* (for *entirely* or *rather*), *rudimentary* (for *vestigial*), *same* (for *identical* or *similar*), *secondly* (for *second*), *singular* (for *notable*), *sometimes* (referring to place instead of time), *superior* (for *better than*), *transpire* (for *happen*), *view* (for *opinion*), *virtually* (for *almost*), *volume* (for *amount*), *wastage* (for *waste*), *weather* (for *climate*), *while* (for *although*), and *whilst* (for *while*). See also Gowers (1986).

Other words commonly misused

The following words, like the measurements recorded by engineers and scientists, should contribute to precision in scientific and technical writing (see page 91).

Approximate(ly) means *very close(ly)* and should not be used if *about* or *roughly* would be better.

Currently means *now*, and the shorter word should be preferred (see Table 6.1), but in most sentences the word *currently* can be deleted (tautology, see Table 6.3). For example, *We are currently . . .* means *We are . . .* (and *We are currently in the process of . . .* also means *We are . . .*).

Hypothesis. If in science something that is not understood (a *problem*) is stated as a question, a hypothesis is a possible answer to that question (a possible solution to the problem) supported by evidence and capable of being tested by an experiment.

Hypotheses, theories and laws are not facts, but attempts to explain or state what seem to be facts. The scientific method depends on the formulation of hypotheses (see page 3). There may be conflicting hypotheses and if one gains general acceptance it may come to be known as a theory (see 'Scientific method', pages 2–3).

Non-scientists may use the words *theory* and *idea* as if they were synonyms but scientists should consider what meaning they wish to convey and use the following words, which are not synonyms, with care: *assumption, conjecture, expectation, fact, guess, hypothesis, idea, impression, law, notion, opinion, presumption, speculation, supposition, surmise, theory* and *view*.

Literally (meaning *actually*) is a word used incorrectly to affirm the truth of an exaggeration, as in 'His eyes were *literally* glued to the television screen'.

Often. People who eat mushrooms *often* die (but people who do not eat them die only once). In the last sentence, and in each of the following extracts, the word *often* is used incorrectly:

'The houses were large in size and *often* inadequately heated.'
This should read: The houses were large, and many were inadequately heated.

'One reason why reports are *often* not well written is . . .'
This should read: One reason many reports are . . .

'People *often* may not know the meaning of words which seem obvious to you.'

This should read: Many people may not understand words familiar to you.

'When people see a word processor for the first time they are *often* amazed.'

This should read: Many people are amazed when they see . . .

The word each of these writers needed to convey the intended meaning was *many*, not *often*.

Progress means a move forward or a change from worse to better, but this word is misused deliberately by many people in attempts to persuade others to accept changes that are clearly not improvements. Indeed, the most outrageous suggestion acquires a certain respectability if someone calls it progress (Orwell, 1946).

Range: the largest and smallest of a sample, or the difference between these measurements.

Refute should be used in the sense of proving falsity or error, not as if it were a synonym for *deny, reject* or *repudiate*.

Significant is a statistical term with a precise meaning, so care is needed in using it in other contexts if readers are to know whether or not you mean statistically significant.

Sophisticated was once an uncomplimentary word implying sophistry and even artfulness but has been over-used to mean *complicated* or to imply, for example, that a new instrument or technique is, in some usually unstated way, an improvement.

Statistics are numerical data systematically collected, and the results of the analysis of such data.

Viable is a term denoting the capacity to live, but in other contexts *not viable* may mean *too expensive* or *will not work*.

Vital means essential to life and should not be used in other contexts.

Many other words have a precise meaning in the language of science – they are technical terms – but in everyday use they have additional meanings (for example: *allergy, neurotic, subliminal*). In scientific and technical writing, care must be taken to use such words in the restricted scientific sense.

Grandiloquence

You may use words that both you and your readers understand, yet write sentences that are difficult to read. For example, long involved sentences with many long words make for hard reading. If you try to impress people by using long words, your studied avoidance of shorter, more appropriate words is more likely to annoy, amuse or confuse than to impress.

This anonymous version of a well-known nursery rhyme pokes fun at grandiloquence:

> Scintillate, scintillate, globule aurific,
> Fain would I fathom thy nature specific,
> Loftily poised in the ether capacious,
> Strongly resembling a gem carbonaceous.

Some people seem to think that scholarly writing must be hard reading, and that a pompous style is necessary to demonstrate their cleverness to the world. Then, the professorial use of pompous language is copied by coteries of like-winded students (Tichy and Fourdrinier, 1988, page 396). In your writing, prefer a short word to a long one (see Table 6.1), unless the long word will serve your purpose better.

Table 6.1 Prefer a short word to a longer word if the short word is more appropriate

Instead of this . . .	prefer this . . .	Instead of this . . .	prefer this . . .
accomplish	do	guidelines	guidance
additional	extra	hypothesise	suggest
anticipate	expect	indication	sign
breakthrough	discovery	initiate	start
commence	begin	modification	change
conjecture	guess	possess	have
consider	think	preventative	preventive
considerable	much	represents	is
construct	build	shortly	soon*
demonstrate	show	subsequently	later
encounter	meet	sufficient	enough
endeavour	try	upon	on
excepting	except	utilisation	use
exhibit	show	within	in
fabricate	build	virtually	almost
firstly	first	currently	now

Note
* Be precise if you can: say when.

Superfluous words

Try not to use two words if only one is needed. In particular, words with only one meaning should never be qualified (see Table 6.2). Facts, for example, are things known to be true (verified past events, things observed and recorded, data). So it is wrong to refer to *the fact that* energy may be involved, to write that the *evidence points to the fact*, or to say that someone has *got the facts wrong*, and to speak of the *actual facts* is to say the same thing twice (tautology, see Table 6.3).

Table 6.2 Words that should not be qualified

Incorrect	Correct
absolutely essential	essential
an actual investigation	an investigation
almost unique	rare
completely surrounded	surrounded
conclusive proof	proof
an essential condition	a condition
hard evidence	evidence
they are in fact	they are
few in number	few
a positive identification	an identification
quite obvious	obvious
real problems	problems
streamlined in appearance	streamlined
quite unique	unique

Table 6.3 Tautology: saying the same thing twice using different words

Incorrect	Correct
every individual one	every one
the reason for this is because	because
reverted back	reverted
related to each other	related
each individual person	each person
in actual fact	in fact
in the rural countryside	in the countryside
a specific example	an example
an integral part	a part
We are currently in the process of	We are
I tentatively suggest	I suggest
completely disappear from sight	disappear
different varieties	varieties
in two equal halves	in halves
symptoms indicative of	symptoms of
or alternatively	alternatively
grouped together	grouped
superimposed over each other	superimposed
percolate down	percolate
eradicate completely	eradicate
give positive encouragement	encourage

Technical terms

In studying any subject we acquire a vocabulary of specialist or technical terms that makes for easy communication between specialists, but which may not be understood by other educated people with different backgrounds and different interests. Before using a technical term, therefore, consider whether or not it will help your readers.

The use of technical terms unnecessarily, or without explanation, may indicate that the writer has not considered the needs of the readers, or not realised that some educated people may not understand the terms – or the words used for something different from their commonly accepted meaning.

Writers who use technical terms after considering their readers' needs make two assumptions that may not be justified: first that readers are familiar with the thing named, and second that they will recognise it by its technical name (Flood, 1957).

Use technical terms when they are needed, not to impress non-technical readers, who are likely to ignore any thoughts, or lack of thought, concealed by a smoke-screen of professional jargon. Wherever possible, replace a technical term by an everyday word if this can be done without altering the meaning of the sentence.

If books, magazines and articles in newspapers intended to popularise science include unnecessary technical terms, and other words that some readers find difficult, they serve as barriers – not bridges – between specialists and other educated people.

If you are writing for non-scientists, or using terms that are defined differently by different people, any necessary terms must be sufficiently explained in simple language. Help your readers by relating a new word to familiar words, by indicating the nature of the thing named, by providing a brief explanation or derivation in parenthesis, by a negative interpolation, or by explaining the concept fully before giving its name (Flood, 1957).

If a technical term is used as a substitute for an explanation, it gives no more than an impression of knowledge (see Beveridge, 1968). For example, the behaviour of an animal may be described as instinctive, but few scientists attempt to define the word *instinct*. Other words that sound like technical terms, but cannot be defined are *libido* in psychology and *ore* in geology. Unless a word can be defined clearly and then used with accuracy and precision, it may conceal our ignorance and obscure the need for further research, and so should have no place in scientific writing.

Many technical terms play an essential part in the prose of science. If they are widely accepted they contribute to an economy of words, and should be part of the common language used by scientists everywhere.

However, other technical terms are short-lived because they serve no useful purpose, or because of misuse, or because they are never clearly defined in an acceptable way, or because they are related to hypotheses and theories that have been superseded. If you express yourself clearly without technical terms you will be understood more widely, and your writing may be understood better by future generations.

If any technical term is to retain its value, scientists must use it correctly – in the same way as other scientists. If there is no internationally accepted definition, they should say whose definition they are following (and give this definition) or they should define the term to make clear their usage. Also, as with other words, the use of any technical term should be consistent throughout any document.

If you do not help non-specialists to understand essential terms, to which they may refer disparagingly as technical jargon, they will not be impressed and will probably lose interest in your message. If any term is essential, you may need to provide a brief explanation when it is first used (or define the term, in the text or in a glossary).

One way of providing a concise explanation is to add a summarising phrase (sign-posted, for example, by the words *That is,* . . . or *That is to say* . . ., or *In short* . . ., or *In other words* . . .) in which everyday words are used instead of the long words or specialist terms that may not be understood by some readers. Such was the habit of Mr Micawber in Charles Dickens' novel *David Copperfield*, written in 1850:

> 'Under the impression . . . that your peregrinations in this metropolis have not as yet been extensive, and that you might have some difficulty in penetrating the arcana of the Modern Babylon in the direction of the City Road – *in short*' said Mr Micawber, in another burst of confidence, 'that you might lose yourself.'

Similarly, in Frank Loesser's lyric for the musical *Guys & Dolls*, written in 1953, Miss Adelaide, a fiancée of many years standing, reads in a technical publication about psychosomatic symptoms affecting the upper respiratory tract and concludes that, *in other words*, she has a cough.

Nomenclature

Scientists would like to give every chemical, every kind of rock, every species of organism, every disease, every part of the body, and every other thing, its own unambiguous and internationally accepted name; and there are international codes for the naming of, for example, animals, bacteria and chemicals.

Trade names

Some words in common use are trade names (for example, Biro, Dictaphone, Hoover and Sellotape) and should therefore have an initial capital letter. However, to make sure you do not misuse them, it is best to avoid trade names if you can. Usually it will also be more accurate and less dated to prefer generic names (for example, ball-point pen, dictating machine, vacuum cleaner and clear tape or masking tape).

Although trade names must sometimes be used, they do not necessarily contribute to accuracy: for example, the chemical composition of a product or the components of an instrument may change although the trade name is unchanged.

Abbreviations, contractions and acronyms

An *abbreviation*, a shortened form of a word, may have several meanings (for example, adv. = advent, advocate, adverb, advertisement; d. = daughter, day, dead, dollar, dose, pence) so even after referring to a dictionary of abbreviations a reader may have to rely on the context in trying to decide which meaning was intended. This is also true of *acronyms*, which comprise the initial letters of successive words and may be pronounced as if they were words: for example, United Nations Educational, Scientific and Cultural Organisation (UNESCO). Furthermore, abbreviations and acronyms in common use in one country may not be understood in another.

So, it is best to avoid abbreviations and acronyms if you can. If any are essential: (a) write them in full where they are first used in any document (each followed immediately by the abbreviation or acronym, in parenthesis – as above); *or* (b) list and explain them at the beginning of a document; *unless* (c) they have come to be accepted as words (as, for example, have the acronyms scuba (self-contained underwater breathing apparatus) and radar (radio detecting and ranging).

In writing English it is also best to avoid phrases from other languages, and abbreviations of such phrases. Any that must be used, if they are not already accepted as English words, should be underlined in handwriting or printed in italics (as in this paragraph: see also page 154). The abbreviations *loc. cit.* (in the place cited), *op. cit.* (in the work cited), and ibid. (in the same work), like the words former and latter, contribute to ambiguity, so these should not be used. Even the abbreviations *i.e.* (*id est* = that is) and *e.g.* (*exempli gratia* = for example) are misused and therefore misunderstood by some people. Write namely (not *viz*) and prefer about or approximately to *circa, ca.* or *c*. The abbreviation *etc.* (*etcetera* = and other things), used at the end of a list, conveys no additional information, except that the list is incomplete. It is better, therefore, to write, *for example* or *including* immediately before the list. These examples illustrate the use of the full stop after an abbreviation.

In *contractions*, which include the first and last letters of a word (for example, Mr, Mrs and Dr), in the letters indicating qualifications (for example, BSc and PhD), and in acronyms (for example, WHO for World Health Organization), full stops are not used (nos., for numbers, is an exception). Also, a full stop should not be used after the symbol for an SI unit (for example, kg and mm: see also pages 93–5) unless this comes at the end of a sentence.

Improve your writing

Take an interest in words

Clearly, it is important (a) that working scientists and engineers should consider carefully the precise terms of reference or the exact meaning of any instruction in relation to any work they undertake; and (b) that students should take an interest in the words used in the questions set in assessed course work and examinations. For example, some students do not answer precisely the question asked because they *either* have not considered the precise meaning of such words as account, brief, concise, criticise, define, essay, *etc.*, or they do not pay enough attention to these words as instructions.

Use your dictionary

Always have a good dictionary to hand, on your bookshelf or in your desk drawer, as a guide to the correct spelling and pronunciation of each word listed, its function, its origin, its current status in the language, and its several meanings.

Choose words with care

1   Cover Table 6.1 with a sheet of paper, then uncover Column 1 and suggest a shorter word that you could use instead of each of the long words, if the shorter word would serve your purpose better. Continue to the end of the table.
2   Cover Table 6.2 with a sheet of paper, then uncover Column 1 and suggest how the same meaning should be conveyed in fewer words.
3   Cover Table 6.3 with a sheet of paper, then uncover Column 1 and, for each entry, suggest which word or words should be deleted.

Define technical terms

A good exercise, to test your understanding of the meaning of a specialist term used in your subject is to attempt to define it, as you would have to do if you were to use it in a composition that was to be read by people who lacked your specialist vocabulary.

Start by listing the points that must be included. Then remember these two rules in writing definitions. First, you must proceed from the general to the particular – from a statement of the general class to which the thing defined belongs, to those features peculiar to the thing defined. Second, your definition must apply to all instances of the thing defined, but to no others. Your definition should also be as simple as possible.

An example (or examples) of the thing defined, although not part of the definition, should be added if this would help the reader to understand.

7 Using words

Unlike the novelist who is trying to paint pictures with words, leaving much to the reader's imagination, your intention in writing about science or engineering is to convey information without decoration: to express your thoughts as clearly and simply as you can.

Words in context

In a dictionary each word is first explained and then used in appropriate contexts to make its several meanings clear. This is necessary because words do not stand alone: each one gives meaning to and takes meaning from the sentence, so that there is more to the whole than might be expected from its parts. It is the function of the words in a sentence to tie one another down so that the sentence as a whole has only one meaning.

The repetition of a word

Some people have favourite words and phrases (for example, also, apparently, case, found, incidentally, in fact, quite). However, the use of a word twice in a sentence, or several times in a paragraph, or many times on one page, may interrupt the smooth flow of language and experienced writers try to avoid such undue repetition. But the so-called elegant variation that results can be overdone. For example, in one paragraph on a sports page of a newspaper a team may be referred to by the club's official name, by the colour of the team's shirts, and by the name of the club's ground. So, a reader has to be familiar with all these names to understand the message.

In scientific writing the right word should not be replaced by a less apt word for the sake of elegant variation. Instead, be consistent: always refer to a spade as a spade. You may also repeat a word to emphasise a point. For example, in the last paragraph the word *by* was used three times in one sentence – to draw attention to each of the items in a list – although only the first *by* was actually needed to make sense.

Words that must be used with care, or ambiguity may result, include: this, that and it; he, him, his, she and her; former and latter; and other and another. If it helps to make your meaning clear at first reading a noun should be repeated.

The position of a word

In a sentence, the position of a word may reflect the emphasis you wish to put upon it. An important word may come near the beginning or near the end, and in either position it may help to link the ideas expressed in successive sentences.

The position of a word may also transform the meaning of a sentence. For example, the word *only* is well known for the trouble it may cause when out of place (see Table 7.1). Consider, also, the meaning of each of the following sentences:

> We only eat fish on Fridays.
> We eat only fish on Fridays.
> We eat fish only on Fridays.
> We eat fish on Fridays only.
> Only we eat fish on Fridays.
> We do not eat meat on Fridays.

The meaning intended in the first sentence is probably that conveyed by the last, which does not include the word *only*. In conversation most people would probably take this meaning, not from what was said but from the context, the intonation and the accompanying facial expression.

Fowler (1968) contradicts himself: stating first that writers should not be forced to spend time considering which part of the sentence is qualified by the word only, and second that it is bad to misplace this word when in the wrong position it would spoil or obscure meaning. Similarly, Gowers (1986) advises that *only*-snoopers should not be taken too seriously – then tells readers to be on the alert. So, be on the alert: try to ensure that an out of place *only* does not spoil or obscure your meaning (see Table 7.1).

If any words in a sentence are misplaced, the meaning conveyed may not be the meaning intended. So, ensure that what you write does express precisely what you mean. Do not expect readers to waste their time trying to guess what you probably meant.

Table 7.1 Only: a word out of place

What the author wrote	Corrected version
The standard only stipulated that . . .	The standard stipulated only that . . .
Some functions can only be performed when online.	Some functions can be performed only when online.
The chemical was only manufactured in Europe.	The chemical was manufactured only in Europe.
In this book those points of grammar only are discussed which will help you to ensure accuracy.	In this book only those points of grammar that will help you to ensure accuracy are discussed.
The census only takes place every ten years.	There is only one census in each decade.
The information is only used for . . .	The information is used solely for . . .

Consider the following sentence from a newspaper:

> Meat Inspectors were reprimanded and downgraded after a consign-
> ment of beef from the local market was shown to be contaminated by
> environmental health officers.

The words *by environmental health officers*, which are out of place, could be inserted after *reprimanded*, or after *downgraded*, or (to give the meaning presumably intended) after *shown*.

Idiomatic expressions

George Orwell (1946), in an essay on Politics and the English Language, complained about the thoughtless use of hackneyed phrases (for example, *with regard to* and *cannot be left out of account*) assembled 'like the sections of a prefabricated hen-house'.

Instead of denying themselves the simple pleasure of putting their own thoughts into their own words, writers should follow Jerry Herman's advice in the musical *Mame* and always 'Open a new door, . . .'

Avoid hackneyed phrases and clichés (and idiomatic expressions, in which the words have a special meaning, see Table 7.2), not only because some readers may misunderstand them, but also because such ready-made phrases make less impact than would a fresh turn of phrase. Instead, choose words that convey your own meaning precisely.

Table 7.2 Some idiomatic expressions

Idiomatic expression	Prefer
break new ground	start something new
leave no stone unturned	make every effort
in the pipeline	in preparation
take on board	note
a different ball game	another matter
see the light at the end of the tunnel	making progress

Circumlocution

A more common fault in writing than the use of the wrong word, or of words in the wrong place in a sentence, is the use of too many words. Although a summarising or qualifying phrase may help the reader (see also *The need for comment words and connecting words*, page 75), any unnecessary words can only confuse, distract and annoy. Also, when too many words are used, time, paper and money are wasted (for example, in word processing, printing and advertising).

In revising any composition, therefore, reconsider each sentence and each paragraph to see if it is necessary, and prune sentences to remove all unnecessary words. Short messages will take less time to type and to read – and should increase your chances of receiving replies that are comprehensive, concise and to the point.

Verbosity

A well-constructed sentence should have neither too many words nor too few; each word should be there for a purpose. A verbose sentence, the result of lack of care in writing or revising, includes extra words that make it more difficult for the writer to convey the meaning intended or to evoke the desired response (see Tables 7.3–7.4). Lack of care in sentence construction may also cause a writer to use hackneyed phrases or clichés in preference to more appropriate words.

In lectures on the art of writing, Quiller-Couch (1916) advised those who would write straightforward prose to prefer concrete nouns (things you can touch and see) to abstract nouns, and listed case, instance, character, nature, condition, persuasion, and degree as examples of abstract nouns that should be used sparingly and with care. Other indicators of jargon are: area, angle, aspect, fact, field, level, process, situation, spectrum, time, and type (see Tables 7.3 and 7.4). Of course there is nothing wrong with any of these words when used to convey meaning.

Table 7.3 Circumlocution: the use of many words where fewer would be better

Circumlocution	Better English
in virtually all sectors of the environment	almost everywhere
on a dawn to dusk basis	from dawn to dusk
on a regular basis	regularly
the reading and learning process	reading and learning
We are in the process of making	We are making
over a period of the order of a decade	for about ten years
during the month of April	in April
on a theoretical level	in theory
on the educational front	in education
I myself would hope	I hope
It consists essentially of two parts . . .	It has two parts . . .
There really is something of an obligation upon us to . . .	We should . . .
Such is by no means the case	This is not so
Most importantly of all, . . .	Most important, . . .
The physical process of writing is . . .	Writing is . . .
The process of revising	Revising
I am in the process of . . .	I am . . .
We are looking to find . . .	We are seeking . . .
An increased appetite was manifested by all the rats	All the rats ate more
In no case did any of the seedlings develop lesions	None of the seedlings developed lesions

Quiller Couch (1916) and Orwell (1946) asked writers who were trying to express rather than to conceal thought: (a) to prefer the short word to the long (see Table 6.1); (b) to prefer the direct word to the circumlocution (see Table 7.4); and (c) to prefer transitive verbs (that strike their object) and to use them in the active voice (see page 74).

Reasons for verbosity

Circumlocution – verbosity – gobbledegook – surplusage – this habit of excess in the use of words, which makes communication more difficult than it should be, is well established in the writing of many educated people. As long ago as 1667, in his *History of the Royal Society*, Thomas Sprat wrote:

> . . . of all the Studies of men, nothing may be sooner obtain'd than this vicious abundance of *Phrase*, this trick of *Metaphors*, this volubility of *Tongue*, which makes so great a noise in the World. But I spend words in vain; for the evil is now so inveterate, that it is hard to know whom

to *blame*, or where to begin to *reform*. We all value one another so much, upon this beautiful deceit; and labour for so long after it, in the years of our education: that we cannot but ever after think kinder of it, than it deserves.

Tautology, circumlocution, ambiguity and verbosity arise from ignorance of the exact meaning of words, from lack of thought when writing, and from lack of care when revising. Also, people may use too few words when they speak, or too many words when they write, if they have not considered the difference between speaking and writing.

In conversation we may use more or fewer words than are needed in writing. On the one hand, we use words to separate important ideas, we repeat things for emphasis, and we correct ourselves in an attempt to achieve greater precision. The extra words give listeners time to think. On the other hand, in conversation we take short cuts, leaving out words, and so use fewer words than would be needed in writing. This is possible

Table 7.4 Circumlocution: some phrases which should not be used if one word would be better

Circumlocution	Prefer	Circumlocution	Prefer
on account of the fact that	as	aimed at	for
if it is assumed that	if	count up	count
a sufficient number of	enough	later on	later
a greater length of time	longer	seal off	seal
during the time that	while	in between	between
it may well be that	perhaps	in regard to	about
using a combination of	from	in order to	to
are found to be in agreement	agree	a proportion of	some*
make an examination of	examine	at a later date	later*
undertake a study of	study	at an early date	soon*
take into consideration	consider	in the nature of	like
it is apparent, therefore, that	hence	it would appear that	apparently
in conjunction with	with	to say nothing of	and
after this has been done	then	has an ability to	can
have a listen	listen	a large number of	many*
have been shown to be	are	for the purpose of	for
for free	free	until such time as	until
carry out experiments	experiment	in connection with	about
come to the conclusion	conclude	located on	on
in view of the foregoing	so	provided that	if
in all other cases	otherwise	spell out in depth	explain

Note
* If possible, be precise. Say how many. Say how much. Say when.

because as we talk we also communicate without words, by a body language in which 'every little movement has a meaning of its own' – and we see when the listener has understood and we have said enough.

The writer, to allow for the lack of direct contact with the reader, must use as many words as are needed to convey the intended meaning. Emphasis is usually made without repetition, and necessary pauses come from punctuation marks and paragraph breaks.

In writing, as in speaking, use words with which you are familiar and try to match your style to the occasion and to the needs of your readers. Write as you would speak to the audience you have in mind, but recognise that good spoken English is not the same as good written English. If a good talk is recorded and then typed verbatim, the reader may find that it is not good prose.

The use of more words than are needed, in writing, may result from a confusion of thought, a failure to take writing seriously, or laziness in sentence construction and revision. All these things are likely when a document is dictated unless it is revised in typescript. Few people are able to dictate anything other than a short routine communication, so that it reads well and conveys the intended meaning, unless they are prepared to spend time converting the typescript into good prose. But most people, if they take the trouble, can write better than they normally talk – because in writing they have more time for thought and the opportunity to revise their work.

Responsibility for revising a typescript cannot be delegated: only the writer knows the meaning intended and whether or not the reader is likely to be affected in the desired way. Before signing any document, therefore, its author must be satisfied with its content and style.

Apart from lack of care, there are other reasons why people fill their writing with empty words. Some seem to think that restatement in longer words is explanation. Others are trying to make a little knowledge go a long way. And others may even be trying to obscure meaning because they have nothing to say, or do not wish to commit themselves:

> . . . only the wealthy, the capable, or the pretty can afford the luxury of saying right out just what they think, and blow the consequences.
> Edgar Wallace (1918) *Lieutenant Bones*

Wordiness may also result from affectation, from the studied avoidance of simplicity, in the belief that Latin phrases, long words and elaborate sentences appear learned (McCartney, 1953). In encouraging direct, straightforward prose, George Orwell (1946) asked writers to be positive, to avoid double negatives (for example, to prefer *possible* to *not unlikely*) and complained about the use of words like *categorical* and *phenomenon* to dress up simple statements and support biased judgements.

In scientific and technical writing, to be read only by specialists, it is not necessary to express complex ideas in language a layman could understand, and it is not necessary to make simple ideas seem complex. Simplicity is the outward sign of clarity of thought. Wordiness is therefore a reflection on a writer's thinking, and a means by which writers conceal their meaning even from themselves.

In an essay *On Style*, Samuel Coleridge (1772–1834) wrote that 'If men would only say what they have to say in plain terms, how much more eloquent they would be'; and Simeon Potter (1966) that 'We shall be effective . . . as writers if we can say clearly, simply, and attractively just what we want to say and nothing more.'

Another cause of verbosity is that some scientists think objectivity is achieved by writing in the passive voice, or they wish to avoid using personal pronouns and write in the passive voice (for example, 'the following results were obtained') when the active voice (for example, 'I observed' or 'we found') would be more direct. Almack (1930) wrote: 'Only in the preface is the first person permitted; the remainder of the thesis should in common decency be written in the third person.' McCartney (1953), however, considered the prejudice against the use of personal pronouns in scholarly writing unwarranted; and Kapp (1973), referring to the need for commenting and connecting words, used the first person freely: 'I must confess, on reading what I have myself written I have frequently caught myself committing the same sin of omission.'

A writer's reluctance to use the first person increases the number of words required, and can make the writing less rather than more objective. *We found* or *I found* communicate something of the excitement of discovery and make clear who was involved. However, never write *we found* when you mean *I found*. The use of the word *we* (for *I*) should be reserved for monarchs, editors and pregnant women.

The first person is to be preferred to such expressions as *it was found that*, which may leave the reader wondering who made the discovery. Similarly, it is not always clear who is meant by *the author* or *the writer*.

The need for comment words and connecting words

A reader's thoughts should move smoothly from each paragraph to the next, but many introductory phrases and connectives can be deleted without altering the meaning of a sentence or disrupting the smooth flow of language. If you omit such superfluous phrases (see Tables 4.1 and 5.1), your writing will be more direct and easier to read – and so be more likely to serve your purpose. See also *Emphasis*, page 82.

Too many words may be used, in a report, in text references to tables and diagrams (for example, the introductory phrases: 'It is clear from a consideration of Table . . . that . . .'; and 'Figure . . . shows that . . .'). These words are superfluous; and they may cause the reader to think that in the table or figure it is necessary to note only one thing. It is better to say whatever you wish to say about the table or figure and then to refer to it by its number (in parenthesis), as in this book. It is also unnecessary in the heading to a table or the legend to a figure, to write: 'Table showing . . .' or 'Figure showing . . .'

However, in practising an economy of words, do not make the mistake of using too few words. In addition to the words needed to convey meaning, include comment words (for example: clearly, even, as expected, and unexpected) and connecting words (for example: first, second, then, therefore, hence, however, on the contrary, moreover, as a result, nevertheless, similarly, so, thus, but, on the one hand, and on the other hand) to help readers follow your train of thought.

Where necessary, provide reminders to ensure the readers always know why what you are saying is relevant to your message. Your message should neither be obscured by a haze of superfluous words nor deprived of words needed to give it strength.

The rule must be to use the number of words needed to convey each thought precisely (without ambiguity), and to ensure that brevity is not achieved at the expense of accuracy, clarity, interest and coherence. In scientific writing clarity and simplicity are not the only considerations (see Chapter 2), but if you intend to be widely understood you will usually want to convey your message as clearly and simply as you can.

Improve your writing

Be clear and concise

Cover Table 7.1 with a sheet of paper, then uncover the first column and for each entry consider where the word *only* should be placed to convey the meaning the author presumably intended.

Cover Table 7.3 with a sheet of paper, then uncover Column 1 and for each entry suggest how the meaning could be better expressed in fewer words.

Cover Table 7.4 with a sheet of paper, then uncover Column 1 and suggest one word that should be preferred to the phrase in each entry. Continue to the end of this table.

Edit the work of others

You will probably find it easier to recognise long words that could be replaced by short words, phrases that could be deleted, and sentences that are verbose, when you read someone else's writing than when you try to revise your own. However, as a result of editing the writing of others, you will start to take more care in revising your own. Each of the following extracts is followed by comments, and a suggestion as to how it could be improved. Cover the comments and suggestions while you consider each extract. Then write your own edited version before you consider mine.

Extract 1

This is to inform you that we have received your manuscript entitled . . . Although we found it interesting, . . .   17 words

COMMENTS

1   The words *This is to inform you* can be omitted without altering the meaning of the sentence.
2   Obviously the manuscript has been received, otherwise there could be no reply.

EDITED VERSION

Thank you for sending your manuscript entitled . . .
We found it interesting, but . . .   12 words

Extract 2

Indeed, it could be said that personal advancement in life lies in the ability to say the right kind of words in the right way at the right time.   29 words

COMMENTS

1   The words 'it could be said that' add nothing to the meaning of this sentence.
2   Personal development must be in life, so the words 'in life' are not needed.
3   Most people would say 'the right things', not 'the right kind of words'.

EDITED VERSION

Indeed, personal advancement depends on the ability to say the right things, in the right way, at the right time.   20 words

Extract 3

People often read instructions only as a last resort, when they can no longer manage without them.   17 words

COMMENTS

1   The first four words convey the opposite of the intended message.
2   The words 'People often' are used when the words 'Many people' are required.
3   The problem is not that people often read instructions, but that many people do not read them at all.

EDITED VERSION

Many people do not read instructions – except as a last resort when they can no longer manage without them.   20 words

Write précis and summaries

When you read, as a student or in other employment, you must recognise the points that are important for your work. These may, for example, be noted on an index card or incorporated in a memorandum.

Preparing a précis is a test of comprehension and an exercise in reduction, in which the essential meaning of a composition is retained – but without ornament and without the details. The order of presentation should not be changed unless it is faulty: the author's meaning should be conveyed in your own words – and in fewer words. As part of a course in scientific or technical writing, a class of students could be asked: (a) to prepare a précis of an article relevant to their studies, working alone; and then (b) to try to agree as to which words in the article can be omitted in the précis.

A *summary*, in scientific writing, should be much shorter than a précis (see page 140). It should include only the author's main points; so preparing a summary is a good test of your ability to recognise these main points and to report them in a few well-chosen words. For practice in preparing a summary, select an article relevant to your own work from a recent issue of a magazine or journal in which authors' summaries are published. Before looking at the author's summary, read the article carefully, listing the main points, and then prepare your own summary (in less than 200 words). Do you agree with the author's choice of the most important points? Has the author used more words than are needed? Have you?

Because it is easier to condense other people's writing than your own, preparing précis and summaries provides practice in recognising important points and will help you to develop a clear, simple and direct style that is appropriate for most scientific and technical writing. So look at each essay or report you prepare to see if by making it shorter you can make it better. An essay is too short to require a summary, but when you have written a report, reconsider your first draft of the summary to check that it does contain each of the points your readers must know about but nothing more.

Write a book review

Many journals and magazines contain reviews of books likely to be of interest to their readers, written by suitably qualified reviewers. A book review is also a useful exercise in comprehension and criticism for students interested in the art of writing. Before writing a book review, read the book. Make brief notes, remembering that to criticise does not mean to find fault with. Criticism of a good book or a good play should be favourable.

The length of the review may be decided by the editor; and if the review is too long, it may be reduced by the editor. The easiest way to do this is to remove sentences at the end – so the most important things must come first and the least important last. The reader needs: title of book (and sub-title); name(s) or author(s) or editor(s) from the title page; date of publication from the title page verso; number of edition (unless the first), name of publisher; place of publication; total number of pages (including preliminary pages); number of tables and figures; and the prices and ISBNs of the hardback and paperback editions.

Readers of a book review have been attracted by the title. They do not want a précis or summary of the book. They do want a brief guide and evaluation, to help them to decide whether or not to look at the book. Answer the following questions: What is the book about, if this is not obvious from the title? Has it any special features? How is the subject treated? What prior knowledge is assumed? For whom is the author writing? Is the treatment comprehensive? Is the book interesting and easy to read? Are the illustrations effective? Is the book well organised? Will the reader, for whom the book is intended, find the book useful? How does the book compare with similar books (if there are any) or with the author's earlier works?

Reviewers who have never written a book are unlikely to appreciate the writer's difficulties. Perhaps this is why some reviewers seem to be looking for the perfect book. Although a reviewer may choose to draw attention to errors if these indicate that the author is not as knowledgeable as he or she should be, it is not the reviewer's task to list every minor fault. Nor is it the purpose of a book review to show that the reviewer is (or is not) clever and witty, and could have written a better book. However, a review should begin or end with the name of the reviewer – who will probably be well-known to readers of the newspaper or journal in which the review is published.

8   Helping your readers

Consider not only what your readers want to know, but also what you need to tell them, by way of explanation or example, to ensure that they understand. Omit anything that is irrelevant, and any unnecessary background information. Only students, who may be expected to display their knowledge, should include details that they expect their readers will already know. At work you are not trying to score marks: you are conveying your knowledge to people who require no more information than will satisfy their immediate needs.

Analysing your audience.   Find out as much as you can about your readers. Consider their age, education, interests and occupations, so that you can anticipate any difficulties – and their likely feelings on reading your message. Some readers may not use English as their first language. Some may be experts in the subject of your composition. Others, although experts in other subjects, may be interested in the possible applications of your work – and be involved in decision-making.

Whatever you are writing, therefore, convey your meaning as clearly and simply as you can, using words, numbers and illustrations, as appropriate, so that all those for whom the document is intended will understand at first reading at least the parts relevant to their work.

Writing for easy reading

Designing your message.   Your writing should be appropriate to the subject, to the needs of your readers and to the occasion. Each sentence should convey a whole thought accurately, clearly and as simply as possible, so that your readers take your meaning and always feel at ease. They are most likely to follow your arguments, understand your evidence, and remember your conclusions, if they can relate anything new to their existing knowledge and interests.

Communicating your purpose. Help readers by providing an informative title, and effective headings and sub-headings. Help them to see the connection between sentences, paragraphs and sections. Sometimes a word is enough; sometimes much more explanation is required.

Obtaining a response. Present information in an appropriate order. Include all essential steps in any argument; give evidence in support of anything new; give examples, and explain why any point is particularly important. No statement should be self-evident, but do not leave your readers to work out any implications. Be as explicit as necessary.

Fulfil your readers' expectations. For example, always follow the words *first* by *second; on the one hand* by *on the other hand; whether* by *or;* and *not only* by *but also.* If you list a number of items, mention all or none of them in the sentences that follow: if only some are mentioned readers may be wondering about the others when they should be thinking about your next topic.

How to begin

If you know what you wish to communicate but have difficulty in getting started, look at the opening sentences in similar compositions by other people. Begin, for example, with: a summary, recommendations, a statement of a problem, a hypothesis, necessary background information that leads directly to the problem or hypothesis, an example, a definition, a question, an answer to one of the readers' six questions (see page 41), an idea that has received some support (then explain why it is incorrect), an accepted procedure (then explain the advantages of an alternative).

The best starting point, for the subject and your readers, will probably be obvious once you have prepared your topic outline. However, it is better to begin than to spend too much time trying to decide how to begin. Your first paragraph can be revised, if necessary, when your first draft of the whole composition is complete. The only rules about beginning are: (a) come straight to the point, with an effective heading or title; and (b) if possible, refer briefly to things you expect your readers to know, and build on this foundation. See also page 42–3.

Control

In each document you write, pay careful attention to presentation – to the arrangement of your material, order and timing – so that you are always in control: communicating information and affecting your readers in a chosen way. Maintaining control depends first on your knowledge and understanding, and then on careful planning – which helps you to present your thoughts in an appropriate, ordered and interesting way. Good headings and sub-headings, especially in a long composition, are signposts that help readers along and – if they are not reading the whole composition – help them to find just the information they require.

Emphasis

The title, headings and sub-headings emphasise the whole and its parts. Emphasis, which is achieved in many ways, is important in all writing and is present whether or not the writer is in control. But you can use emphasis effectively only if you know how to make important points stand out from the necessary supporting detail.

Beginnings and endings are important. The first and last paragraphs (the introduction and conclusion) will be read by most people. Then in each paragraph the first and last words capture most attention. In planning a composition you have to decide on the order of paragraphs, and you may number them in your topic outline. But remember that your plan is for you, not for the reader who requires only the results of your thinking and planning.

So, omit such superfluous introductory phrases as: *First let us consider . . .; Secondly it must be noted that . . .; An interesting example which should be mentioned in this context is . . .; Next it must be noted that . . .; In conclusion it must be emphasised that . . .* Also, omit other unnecessary introductory phrases and connecting phrases (see Tables 4.1 and 5.1).

Never begin a paragraph with unimportant words; and end each paragraph effectively. Similarly, in a sentence emphasis falls naturally on the first and last few words: so use these words to convey information or to make connections – to help readers understand your message and follow your train of thought.

A reader's or listener's attention can be captured and held by saying things in threes: a technique over-used by some politicians. It is no accident that in ancient times there were three Graces, and in the Christmas story three wise men. Saying things in threes encourages the reader or listener to anticipate what is to be said next and makes it easy to remember what has been said.

Items of comparable importance can be emphasised by repeating an introductory word, by numbering, or by indentation. However, if a sentence has been properly constructed, so that it reads well, emphasis will fall naturally on each part. Similarly, if a composition has been well planned it will be well balanced, with an obvious beginning, middle and end, and each paragraph break will serve to emphasise that one topic has been dealt with and it is time to start thinking about the next.

If appropriate, plan effective illustrations to convey the essential points. In writing, use more forceful language for important points than for any supporting detail; and check your first draft to ensure you have emphasised them sufficiently. In your topic outline you may underline words or phrases to remind you of points you intend to emphasise in your composition, but in the composition itself do not underline for emphasis. Underline only those words that in a book or journal would be printed in italics (see page 154).

Sentence length

Long involved sentences may indicate that you have not thought sufficiently about what you are trying to say. If as you revise your composition you find a long sentence that is difficult to read, consider how it can be improved. Perhaps it should be broken into two or more shorter sentences.

The breaks between paragraphs and sentences give readers time for thought; and in a newspaper the length of paragraphs, sentences and words is intended to match what the editor thinks are the readers' needs. In some newspapers each paragraph is one short sentence. In others the paragraphs are longer, some sentences are longer, and a wider vocabulary is used.

However, although short sentences are the easiest to read, a long sentence, if it is properly constructed, may be easier to read than a succession of short ones. There is no rule that a sentence, when read aloud, should be read in one breath. Good prose is seldom written in short sentences. An opinion can be clearly expressed, even in a long sentence, as in the following 48-word sentence from a novel:

> It is the fashion now 'to go along with the people' but I think the people ought to be led, ought to have ideas given them by those whom nature and education have qualified to govern states and regulate the conduct of mankind.
>
> Disraeli, B. and Disraeli, S. (1834)
> *A Year at Hartlebury or The Election*

Sentences vary in length. Short sentences are effective for introducing a new subject, long sentences for developing a point, and short sentences for bringing things to a striking conclusion, as in this extract from another novel:

> 'If you really want to know,' said Mr. Shaw, with a sly twinkle, 'I think that he who was so willing and able to prove that what was was not, would be equally able and willing to make a case for thinking that what was not was, if it suited his purpose.' Ernest was very much taken aback.
>
> Samuel Butler (1903) *The Way of All Flesh*

Rhythm

Good prose, like speech, has a varied rhythm that contributes to the smooth flow of words in a sentence, gives emphasis to important points, and makes for easy reading. In contrast, badly constructed sentences may irritate readers and make them less receptive to your message. So it is a good idea to read your writing aloud, and to revise any parts that do not sound well.

McCartney (1953), in *Recurrent Maladies in Scholarly Writing*, asks writers to be sensitive to the sounds of words and to try not to offend the ear, for example: (a) by unintentional alliteration, as in *rather regularly radial*; (b) by the grating repetition of s, as in *such a sense of success*; (c) by adding s to a word that does not require it, for example to *forward and toward* [but the s may be needed to make the sentence easier to read]; (d) by the repetition of syllables, as in *appropriate approach*, *continue to contain*, and *protection in connection with infection*; (e) by the repetition of sound, as in *found around*, and *with respect to the effect*; (f) by the repetition of cognate forms in different parts of speech, as in *a locality located*, *the following procedure should be followed*, *except for rare exceptions*, *no real realisation*; or (g) by repeating a word with a change in meaning, as in *a point to point out*.

Style

Some may feel that style is not important in scientific and technical writing; but style is not something that can be added to writing as a final polish. It is part of effective prose. Graves and Hodge (1947), in *The Reader Over Your Shoulder*, emphasised: (a) clarity, completeness, consistency, order, simplicity, sincerity and consideration for the reader as basic requirements; (b) that all connections should be properly made; and (c) that although written for silent reading, effective prose should sound well if read aloud.

The importance of planning was emphasised by George de Bufon, addressing the *Académie Française* in 1703: 'This plan is not indeed the style, but it is the foundation; it supports the style, directs it, governs its movement, . . . Style is but the order and the movement that one gives to one's thoughts.'

Because the way you put words together reflects your own personality and your feeling for words, it would be a mistake to try to copy someone else's style. In writing about science, a good style depends upon your intelligence, imagination and good taste; on sincerity, modesty, careful planning, and attention to the requirements of scientific writing (see pages 30–5). You are familiar with these things as part of the scientific method. In effective prose the excitement of discovery may also be communicated.

Rhythm, while not essential, will make for easier reading, and badly constructed sentences may irritate readers and make them less receptive to your message.

Capturing and holding your readers' interest

In writing about science or engineering in a book, in an article for a journal, in a project report, or in describing an experiment, your interest in your subject should be conveyed to your readers.

A novelist, whose business is words, takes great care over the choice and use of words. Consider, for example, how in the first paragraph of a good novel the author captures the readers' interest and begins to tell the story. You will find no superfluous words. In writing about science you start with the advantage that your readers are already interested, but to maintain their interest you must present information at a proper pace. If readers understand, they will want to move quickly to the point; but they must understand every word, every statement and every step in any argument. If they have to refer to a dictionary, or read a sentence more than once, before they can understand your message, you may lose their attention.

Readers are directed away from an explanation or argument by anything irrelevant, by unnecessary detail, by explanation of the obvious, and by needless repetition. They lose interest if statements are not supported, as appropriate, by evidence or by examples. Science depends upon evidence and you must not attempt to gain acceptance of your views by reiteration. Use cross-references to avoid repetition and to provide necessary reminders. When you repeat anything deliberately, using different words, either for emphasis or to help to clarify a difficult point, use a phrase such as *That is to say* or *In other words*. Otherwise, after studying both sentences, readers may be left wondering if they have failed to appreciate some difference in meaning. See also *Technical terms*, page 62.

Approach people through their interests rather than your own. They will be most interested in themselves, in other people and in things as they affect people. Scientists will be most interested in their own speciality and in developments likely to have a bearing on their own work. Other people are likely to be interested in the application of science and technology to human welfare, in the impact of discoveries on society, and in pure research when this is concerned with our origins and our place in the universe. See also *How to begin*, page 81).

In an internal report or journal article the style of writing is usually direct and the link between paragraphs is achieved mainly by their orderly arrangement. In a magazine with a wider readership more explanation and interpretation is needed; and in a newspaper attention is maintained by reference to familiar things, by including examples, anecdotes and analogies, and by providing attractive illustrations.

For an even wider audience, a sign (for example a traffic sign), a cartoon (as used in the popular press to highlight the day's main story), or two or more drawings (as in a strip cartoon, see Figure 8.1), can be used to capture attention – including that of people who would not – or could not – read written instructions.

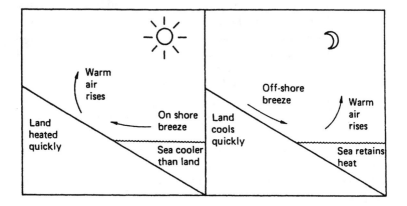

daytime sea breeze; night-time land breeze

Figure 8.1 Drawings used, as in a strip cartoon, to help explain the causes of sea breezes. Diagram from Barrass, R. (1991) *Science*, Basingstoke, Macmillan.

Using good English

Looking critically at other people's writing will help you to improve your own, but do not be afraid to put pen to paper for fear of making mistakes. English is bad only if it does not express the thought intended clearly and accurately in words appropriate to the context. However, even if a sentence is grammatically correct, superfluous words make for hard reading (see Table 8.1). In scientific and technical writing clarity depends on the use of words readers will understand and expressing thoughts as simply as you can.

Table 8.1 Advice on the use of language* (and a shorter version conveying the same information)

Extract	Précis
The purpose of these general notes is first to assist authors in the writing of scientific papers in an acceptable style and secondly to suggest the introduction of greater uniformity in the general approach to the preparation of scientific papers for publication. It is, nevertheless, clear that specialised subjects may call for particular methods of presentation, and these notes should be read in conjunction with any instructions issued by the journal in which the author hopes to publish. When the notes conflict with such instructions, the latter should be followed. (90 words)	These notes are to help authors write scientific papers in an acceptable style, and to encourage greater uniformity in presentation. But, if these notes conflict with the editorial policy of the journal in which you hope to publish, the rules of the journal must be followed. (46 words)

Note
* Extract from Royal Society (1974) *General Notes on the Preparation of Scientific Papers*, 3rd edition, London, Royal Society. For more extracts from published and unpublished compositions, and suggested improvements, see Gowers, E. (1986) *The Complete Plain Words*, 3rd edition, London, HMSO, Ch 3 'The Elements'.

Poor writing may result from distraction, from not knowing what to say, from not considering how to present information, from insufficient care in the choice and use of words, or from not allocating sufficient time to thinking, to planning, to writing, and to checking and if necessary to revising the work. Poor writing is also to be expected from a writer who has nothing to say, or who does not wish to express an opinion, and is so inconsiderate as to try to put up a smoke-screen of words that gives the impression that something is being said, but only obscures meaning.

Obstacles to effective communication

Communication is not easy: an effort is needed on the part of the writer if the reader is to be interested, informed and affected in a chosen way (see *Appropriateness*, page 35). Failures in written communication between educated people may result, for example, from: (a) lack of practice on the part of the writer; (b) the writer's unwillingness to devote enough time to thinking, planning, writing and revising; (c) failure to establish contact with readers at the start; (d) lack of attention on the part of readers, especially when the writing deviates from their interests; (e) the readers' preconceived ideas, and their refusal to accept new ideas or to consider evidence that conflicts with their existing beliefs.

Rules for efficient communication

To write well most people need to be alone, free from disturbance, and with time for thought (see *Writing*, page 44).

1 Before starting to write, decide whom you hope to interest, why you wish to interest them, what must be said, and how you should say it. Readers are most likely to understand grammatically correct English, and be interested in evidence (summarised in tables and effective illustrations).
2 Write about things you know, if you have something interesting to say.
3 Plan your work so that information and ideas can be presented in an appropriate order, and so that the whole composition has the qualities of balance and unity.
4 Write for easy reading. Begin well. Keep to the point. Be clear, direct and forceful. Maintain the momentum of your writing, if possible by writing at one sitting.
5 Check your work, and revise it if necessary.

Improve your writing

Learn from people who write well

In starting to play any game you can learn much by watching experts. Similarly, reading good prose will influence the way you write, just as the way you speak is influenced by the speech you hear.

Read books by successful authors and study the techniques of journalists who write well – to see how to capture attention, how to match your writing to the needs of your readers, and how to write clear, concise, vigorous and vivid prose. Consider, for example, the purpose and scope of a leading article in a newspaper, or an article that interests you in a magazine. The title captured your interest. Does the opening sentence make you want to read on? Try to reconstruct the author's topic outline by picking out the topic for each paragraph. Is each paragraph relevant to the title? Are the paragraphs arranged in an appropriate order? Do they lead to an effective conclusion?

Study one paragraph. Note the ideas presented in each sentence. Which is the topic sentence? Are all these ideas relevant to the topic? Why are they presented in this order? Is it the most effective order in helping the writer to make a point (in helping the reader to understand)? Can you distinguish facts from opinions? Are the opinions supported by evidence? Is the article biased in favour of a particular point of view?

What is published in a newspaper is likely to be well written and persuasive, and to interest the people who normally read that paper, but if you compare accounts of one event in different newspapers you will probably find that they tell very different stories. This is because eye-witnesses of one event see and remember different things, and are influenced by their own previous experiences. Then, the stories submitted by reporters are edited to fit the space available in the paper, to match the readers' interests, and to suit editorial policy.

A simple, straightforward style is required in scientific and technical writing; and young people who are still developing a style of their own will find such direct prose in books by, for example, Daniel Defoe, Jonathan Swift, Robert Louis Stephenson, John Buchan, Aldous Huxley, Jack Schaefer and Winston Churchill.

As you study the writing of others, and consider how your own writing can be improved, remember that there is no one correct way to write. Read for pleasure, not to copy someone else's style (see *Style*, pages 84–5). The way you write should reflect your personality and your feeling for words.

Learn by writing

Evans (1972) includes editing exercises, each with three versions of a news item. Version A: the story as it appeared in print. Version B: the story edited to remove superfluous words and improve the English. Version C: rearranged and re-written to bring out the human interest.

Most people can improve their writing by considering the advice of more experienced writers, and from colleagues willing to read and comment on their work, but the best way to learn is by writing. Think before you write, plan your work, try to write without interruption, check your work carefully, and revise each composition until you are satisfied that it will serve your purpose (see pages 40–9). Your writing will improve.

Check your writing for readability

Flesch (1962), in *The Art of Plain Talk*, graded writing simply, according to average sentence length, as very easy to read (less than 10 words), difficult (more than 20 words) and very difficult (30 words). Accepting this as a rough guide to readability, it is worth calculating the average sentence length in a few paragraphs of a document you have written recently. When writing documents that will be read only by your colleagues, you may know they all can cope with long sentences that some people would find difficult to read. But remember, when writing for a wider audience, that some people have difficulty with even short words in short sentences. So, prefer a short word to a longer word if the short word will serve your purpose (see Table 6.1); and try to ensure that every sentence is carefully constructed, grammatically correct and easy to read. Try to convey your message as clearly and simply as you can.

With a word count, using a word processor, it is easy to estimate the average length of words used in a document (by dividing the number of characters by the number of words); and Pullin (2001) suggests that an average of four to six is satisfactory, but that the nearer the score is to four the more readable the document is likely to be.

9 Numbers contribute to precision

A politician may say that a fund will be established, 'of *substantial* size, and *adequate* coverage over a *considerable* period'. Such vague words are used to express hopes when it is impossible to be either certain or precise. In scientific and technical writing consider the meaning you wish to convey: (1) before using the word *very* with an adverb (for example, very *quickly*) or with an adjective (for example, very *large*); (2) before using adverbs (for example, *slowly*) or adjectives (for example, *small*, *appreciable*, *large* and *heavy*); and (3) before using modifying and intensifying words (for example, *comparatively*, *exceptionally*, *extremely*, *fairly*, *quite*, *rather*, *really*, *relatively*, and *unduly*). Such meaningless modifiers result in vague statements that do not help your readers, and may even annoy them:

> Whenever anyone says I can do something soon I'll say to them yes, I know all about that . . ., but when, when, when?
>
> Alan Sillitoe (1961) *Key to the Door*

Using numbers and SI units

Numbers

Be precise whenever you can. Use numbers to make clear how many, and use numbers and SI units of measurement to make clear, for example, how far, how long, how much, how thick.

Use arabic numerals, not words, for the number of the year, but name the month to make misinterpretation of a date impossible: write 4 July 2010, not 04.07.2010 and not 07.04.2010.

Roman numerals are used with the names of monarchs (for example, Queen Elizabeth II), but they are no longer used for numbering photographs or tables – so they have no use in scientific or technical writing.

Where practicable, prefixes and symbols should be used to indicate decimal multiples and submultiples (see Table 9.1) and prefixes involving powers of three are to be preferred. Because of possible confusion arising from differences in usage in different parts of the world, the words billion, trillion and quadrillion should not be used.

Table 9.1 Multiples and submultiples

Prefixes and symbols used with SI units to indicate decimal multiples and submultiples. Prefixes involving powers of three to be preferred.

Multiples			Submultiples		
Factor	*Prefix*	*Symbol*	*Factor*	*Prefix*	*Symbol*
10^{18}	exa	E	10^{-1}	deci	d
10^{15}	peta	P	10^{-2}	centi	c
10^{12}	tera	T	10^{-3}	milli	m
10^{9}	giga	G	10^{-6}	micro	μ
10^{6}	mega	M	10^{-9}	nano	n
10^{3}	kilo	k	10^{-12}	pico	p
10^{2}	hecto	h	10^{-15}	femto	f
10	deca	da	10^{-18}	atto	a

In writing, cardinal numbers (twenty-one to ninety-nine) and ordinal numbers (for example: twenty-first, one-hundred-and-first) should be hyphenated.

Use words, not figures, if a number is necessary at the beginning of a sentence. Use words for numbers one to nine, except before a symbol (six metres, but in technical writing 6 m) or before a percentage sign (6% in a table or figure, but 6 per cent in the text). Prefer figures to words if different items are listed in the same sentence.

Do not write two numbers together, either as figures or words, because ambiguity may result: write two 50 W lamps, not 2 50 W lamps and not two fifty watt lamps.

Where necessary, percentages should be defined (for example, in describing solutions percentage by mass must be distinguished from percentage by volume).

Decimals are indicated by a full stop on the line in some countries and by a comma in others, not by a point raised above the line. So, to avoid confusion with the decimal point, numbers up to 9999 should be without a comma. Instead, spaces should be left if there are more than four digits above or below the decimal point (for example: write 999 999 and 10 000 but 9999), unless it is necessary to leave a space so that in a list the numbers, decimal points and spaces are in vertical alignment (which explains why there is a space between the 8 and the 5 in the number 8 537 in Table 9.3 but no space in the other four figure numbers in this table).

The use of decimals should contribute to accuracy in measurement. The numbers 5 and 5.0 and 5.00 indicate different degrees of precision. For values less than one a zero should be placed before the decimal point (0.25 not .25). Modifying words such as about, more and less should not be used with decimals. Remember that the results of a calculation should not be expressed in more places of decimals than are present in the least accurately known component of the calculation; otherwise your results will appear to be more precise than is possible with the method of measurement used in obtaining the data. Remember, also, that accuracy in calculation can do nothing to compensate for lack of care in the collection and recording of data.

Original data are not usually presented in the body of a report, but are summarised in graphs, diagrams or tables, or described by statistics in the text. The methods used in statistical analyses should be indicated. The symbols should be those used in a recent and authoritative book on statistics, and their meaning should be explained.

Differences which are not statistically significant should not be described as insignificant, and scientists interpreting the results of statistical analyses should remember that when something is improbable this does not mean that it will never happen. When things occur in sequence, the first is not necessarily the cause of the second; and when two things are shown to be correlated, this does not mean that one is necessarily the cause of the other.

SI units

Most countries have adopted the metric system of measurement and use the International System of Units (Système International d'Unités, abbreviated to SI units (see Table 9.2 and the footnote).

The magnitude of any physical quantity must always be stated as the product of a pure number and an SI unit (physical quantity = number × unit). When using symbols, instead of words, the following rules apply.

1 Leave a space between the number and the symbol (for example, 50 W and 20 °C).
2 Do not put a full stop after the symbol unless it comes at the end of a sentence.
3 Do not add an s to any symbol: with SI units the same symbol is used for both singular and plural (m = metre or metres).
4 Do not leave a space between a prefix and a symbol: ms = millisecond.

Table 9.2 International System of Units (SI units)*

Quantity	Unit	Symbol
length	millimetre (0.001 m)	mm
	centimetre (0.01 m)	cm
	metre	m
	kilometre (1000 m)	km
area	square centimetre	cm^2
	square metre	m^2
	hectare	ha
volume	cubic centimetre	cm^3
	cubic metre	m^3
capacity	millilitre(0.001 l)	ml
	litre	l
mass	gramme (0.001 kg)	g
	kilogramme	kg
	tonne (1000 kg)	t
density	kilogramme per cubic metre	kg/m^3
time	second	s
	minute (60 s)	min
	hour (3600 s)	h
	day (86 400 s)	d
speed, velocity	metre per second	m/s
	kilometre per second	km/s
plane angle	radian	rad
solid angle	steradian	sr
frequency	hertz	Hz
force	newton	N
pressure	pascal	Pa
energy, work, quantity of heat	joule	J
electric current	ampere	A
power, energy flux	watt	W
	kilowatt	kW
electric charge	coulomb	C
electric potential	volt	V
electric resistance	ohm	Ω
electric conductance	siemens	S
electric capacitance	farad	F
magnetic flux	weber	Wb
magnetic flux density	tesla	T
inductane	henry	H
luminous flux	lumen	lm
illuminance	lux	lx
luminous intensity	candela	cd
luminance	candela per square metrre	cd/m^2
thermodynamic temp. (T)	kelvin	K
temperature (t)	degree Celsius	°C
amount of substance	mole	mol
concentration	moles per cubic metre	mol/m^3

Notes to Table 9.2*
In the International System of Units, the metre, kilogramme, second, ampere, kelvin, candela and mole, are *base units*. Other units, such as the centimetre and kilometre, are *derived units*, being submultiples or multiples of base units. The radian and steradian are supplementary units (not classified as either base units or derived units). The litre, tonne, minute, hour, day, and the degree Celsius (but not the micron) are recognised units outside the International System. The hectare is accepted temporarily in view of existing practice. In Britain the degree Celsius used to be called the degree Centigrade. For further information on SI units, including units not shown in this table, see list of Standards on pp. 121.

5   Leave a space between the symbols when two or more symbols are combined to indicate a derived unit: metres per second = m s^{-1} (or m/s). Acceleration is indicated as m/s^2 (not as m/s/s).

6   Do not leave a space between the degree sign and the letter C, but do leave a space between the degree sign and the preceding numeral: 20 °C, not 20°C and not 20° C

7   Symbols for physical quantities are printed in italics. Symbols for units are printed in roman type. If, on a graph, potential difference (V) measured in volts (V) is to be plotted against current (I) in milliamps (mA), the axes should be labelled:
   either V/V and I/mA
   or V in volts and I in milliamps

8   Symbols for vector quantities are printed in bold face italic type (for example, F for force). Symbols for tensors of the second order should be printed in bold face sans serif italic type (for example, S).

Preparing tables

In a table, words and numbers are arranged in columns for easy reference. The tables in a document should be numbered consecutively. The size of each table should be decided in relation to the size and shape of the page or column in which it is to fit. If possible, it should fit upright on the page (portrait, not landscape) so that readers can look from the text to the tables without having to rotate the document. Each table should be on a separate sheet with a clear and concise heading above the table (see Table 9.2), and should include sub-headings if they will help the reader.

It should be possible to understand the tables without reading the text, but there should be at least one cross-reference to each table in the text (as in the last sentence). The information provided in a table should not be repeated in the text, or in an illustration, and a table should not include columns of numbers if these could be calculated easily from numbers in other columns.

Table 9.3 The world: population and land

World region	Population* (millions)		Surface area (000s km²)
	1950	2000	
Africa	224	836	30 306
North America	166	308	21 517
Latin America	166	527	20 533
Asia	1403	3744	31 764
Europe	549	734	22 986
Oceania	13	31	8 537
World totals	2520	6169	135 641

Note
* Based on data from UN (1997) *Statistical Yearbook*, New York, United Nations.
The population estimates for 2000 calculated on the assumption that the annual rate of increase from 1995 to 2000 was the same as from 1990 to 1995.

The first column on the left of a table is called the stub. This labels the horizontal rows of the table, indicating what the investigator has decided to study (called the independent variable). In a table used to record numerical data or the results of the analysis of such data, the stub could, for example, state the times at which readings were taken, the names of individuals or nations, or (as in Table 9.3) the world regions selected for study.

The data or results recorded in other columns of a table, the values of which will depend on changes in the independent variable, indicate changes in dependent variables. There is one column for each dependent variable studied, as indicated by concise column headings – which must include units of measurement for every quantity shown (and, as appropriate, a prefix: see Table 9.1). If there is no entry in any part of a table, this should be shown by three dots . . . and a footnote stating that no information is available. A nought should be used only for a zero reading.

In a table comparison should be possible both vertically and horizontally; and where a total is given in the bottom right-hand corner, the vertical and horizontal totals must agree. If a table is to be published, consult a recent issue of the journal or the publisher's house rules for advice on the use of horizontal and vertical ruled lines.

Any necessary footnotes should be immediately below the table to which they apply, but in a handwritten or word-processed document there should normally be no other writing on the same page. Each footnote should be preceded by a superscript reference letter or symbol (see Tables 3.1 and 9.3), not by a number. These reference letters or symbols must also be included in the table, in superscript, to identify the entries to which the footnotes refer (as in Tables 3.1 and 9.3).

If tables of data are necessary, for example in a report, these are best placed in an appendix so that they are readily available for reference but do not distract the readers' attention from your argument in the text. In contrast, most text tables should be concise summaries (results of the analysis of data), to provide readers with just the information they need and to help you to make a point.

Whether tables are in the text or in an appendix, horizontal and vertical ruled lines should be included only if they will help readers. In most text tables vertical ruled lines are not necessary; and the parts of a table can be indicated by concise sub-headings or separated by leaving extra space between horizontal rows rather than by horizontal ruled lines.

Preparing graphs and charts

Types of diagram used for presenting numerical data, or results of the analysis of such data, include the line graph or line chart, the histogram, the vertical bar chart, the horizontal bar chart, the pictorial bar chart and the circular chart (also called a pie chart).

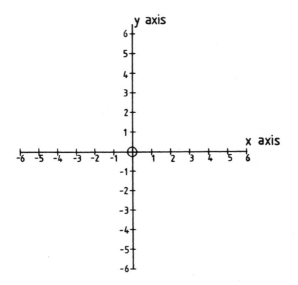

Figure 9.1 The parts of a graph. The thing that you can control must be plotted in relation to the *x*.

Line graphs

Numerical data or results derived by the analysis of data may be presented on a graph in such a way as to facilitate comparison. The points on a graph may therefore be single measurements (in a scatter diagram) or they may be average values. If the latter, either the standard error ($\pm S\bar{x}$) or the 5 per cent fiducial limits of error in the mean ($\pm 1.96\ S\bar{x}$) may be shown by a vertical line through the mean, with a note to indicate which is represented.

There are conventions to be followed in the preparation of graphs (see also, page 113). In a two dimensional graph the horizontal axis (the x axis or abscissa) and the vertical axis (the y axis or ordinate) cross at right angles. On the x axis, positive values are shown on the right of the y axis and negative values on the left. On the y axis, positive values are shown above the x axis and negative values below (see Figure 9.1). Each point on the graph is described by two numbers, its coordinates, which give its position with respect first to the x axis and then to the y axis.

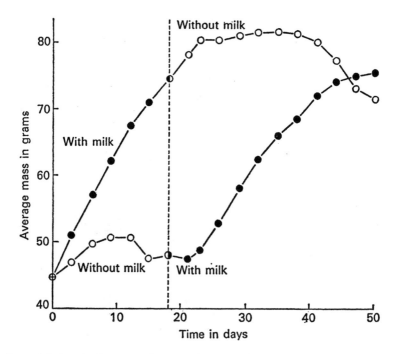

Figure 9.2 Graph: changes in the mass of rats on a diet of pure protein, fat, carbohydrate, mineral salts and water, but either with or without milk supplement. Based on data from Hopkins, F. G. *Journal of Physiology*, 44, 1912.

A line graph shows how one thing varies relative to changes in another. The variable identified by the investigator (for example, in Figure 9.2, the time in days at which average mass is to be recorded) is called the independent variable and must be plotted in relation to the horizontal axis (the x axis). The other variable, which the investigator cannot decide in advance (for example, the average mass), and which depends on changes in the independent variable, is called the dependent variable and is plotted in relation to the vertical axis (the y axis). Only pure numbers are plotted, and points on the graph are marked by symbols.

The scales for the axes of a graph should normally start from zero: they should be chosen carefully and marked clearly. If it is impracticable to start the scale from zero, the break in the axis should be clearly indicated by a jagged line. All numbers should be upright but the labelling of the scales should be parallel to the axes (as in Figure 9.2). Units of measurement must be stated. The diagram as a whole is the graph (line chart) and the lines on the graph, representing trends, even if they are best-fitting straight lines, are called curves (as seen in Figure 9.2 where there are two curves: one for rats with milk, and one for those without).

Joining the points on a graph by lines is called interpolation; and continuing a line beyond the points on a graph is called extrapolation. Extra readings between and beyond the points would not necessarily fall on the lines. That is to say, both interpolation and extrapolation are speculation, which may mislead the writer as well as the reader. A remark by Winston Churchill, made in another context, is appropriate: 'It is wise to look ahead but foolish to look farther than you can see.'

Histograms

A histogram can be used to represent a frequency distribution in which the variation in the data is continuous (meaning that the observations recorded do not fall into distinct or discrete groups). As in a graph, the independent variable being studied (for example, in Figure 9.3, the heights of people) is plotted in relation to the horizontal axis: the number on the left of each vertical column indicates the lowest measurement included in that grouping interval. The vertical column for each grouping interval shows the frequency of observations in that interval. Adjacent columns touch, indicating that the variation is continuous. Note that the scale on the vertical axis starts from zero. As with a graph, all numbers should be upright but the labelling of the scales should be parallel to the axes (as in Figure 9.3).

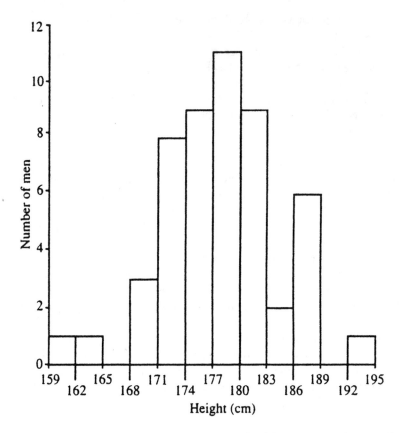

Figure 9.3 Histogram: heights of fifty-one men aged eighteen to twenty-five. Based on data from Harris, A. (1978) *Human Measurement*, London, Heinemann.

Bar charts

A *vertical bar chart*, also called a column graph or column chart, can be used to represent a frequency distribution in which the variation in the data is discontinuous (the observations recorded do fall into discrete groups). As with line graphs and histograms, the variable being studied (the independent variable) is plotted in relation to the horizontal axis, and the length of a vertical column or bar indicates the frequency of observations in each group (the number of a dependent variable at different times or under different conditions). See Figure 9.4(a)

Adjacent columns should be labelled separately and should not touch, emphasising that the variation is discontinuous. As the data are discrete,

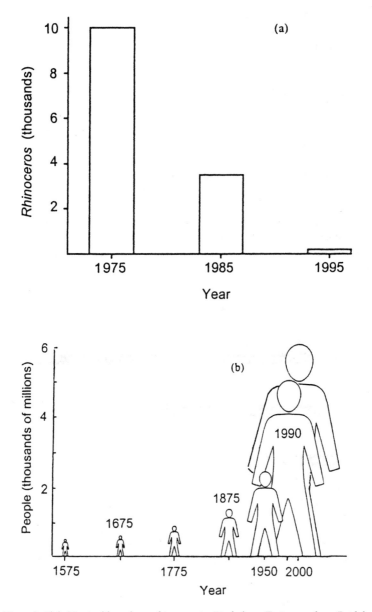

Figure 9.4(a) Vertical bar chart: *rhinoceros* in Zimbabwe. Estimates from Bridgland, 'The end of the rhino'. *Sunday Times Review*, 11 December 1994. *(b)* Pictogram: the growth of world population 1575–2000. Note that the population size is represented by the height of each symbol, not by its area (see text). Estimates for 1950, 1990 and 2000 based on UN (1997) *Statistical Yearbook*, New York, United Nations.

there is no difficulty in assigning each observation to one group. For example, the number of children in a family or the number of rhinoceros in a locality, must be a whole number.

The columns must be rectangles (as in Figure 9.4(a)) because it is the height of a column, not its area, that corresponds to the quantity represented. Drawings should not be used instead of columns (as in Figure 9.4(b)) because differences in the area of the drawings are likely to mislead readers.

Readers may also be misled if a scale on a graph, histogram or other kind of chart does not start from zero. The zero is said to be suppressed or false, and this can make a small difference appear greater than it actually is. Some readers may consider an illustration with a suppressed zero, or with an otherwise inappropriate scale, to be a deliberate attempt to mislead them.

In non-technical writing a column chart may be drawn on its side (as a *horizontal bar chart*, with the dependent variable represented on the horizontal axis) if horizontal bars are more appropriate, make more impact, and so help to convey a message more effectively (see Figure 9.5).

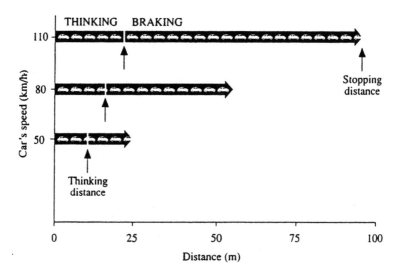

Figure 9.5 Horizontal bar chart: stopping distances. These are the shortest stopping distances for alert drivers of cars with good brakes and tyres, on dry roads, when travelling at different speeds. Speed, decided by the driver, should be plotted in relation to the horizontal axis of a graph, but this chart is drawn on its side for visual effect – and greater impact. Distances are shown in metres and also in car lengths. Based on data from *The Highway Code*, London, HMSO.

In a *pictorial bar chart* (see Figure 9.6) the bars are replaced by identical symbols. Note that, as in this illustration, a bar chart can be used to show how one or more things vary in relation to another when one of the variables is geographical or qualitative (not numerical): for example, (a) in Figure 9.6, the different world regions; and (b) in a wall chart, used as a year planner, different activities scheduled for different weeks.

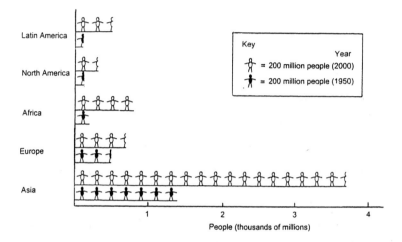

Figure 9.6 Pictorial bar chart: population growth in different world regions between 1950 and 2000 (based on data presented in Table 9.3). Note that Oceania (Australia, New Zealand, and the Pacific Islands excluding Hawaii) cannot be represented because (see Table 9.3) only one-eighth of a symbol would be required.

Pie charts

In a pie chart (also known as a sector chart, circle chart or circular graph) slices of the pie (sectors) are arranged in order according to their size, clockwise, starting at noon with the largest slice (and each slice representing a fraction of 360 degrees). For example, if 836 million people were living in Africa in the year 2000 and the world population was 6169 million, a 49° sector of the chart can be used to represent the population of Africa in 2000 (calculated by dividing 836 by 6169 and then multiplying the result by 360).

If two pie charts are to be compared, the slices in the second should be arranged in the same order as in the first (as in Figure 9.7). The difference in the area of the two charts in Figure 9.7 represents the difference in world population.

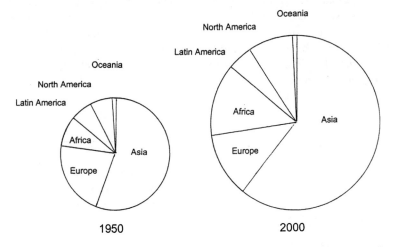

Figure 9.7 Circular graph or pie chart: where people lived in 1950 and in 2000. The difference in area in these two charts represents the fact that the world population more than doubled in these 50 years. For a more informative presentation of these estimates see Table 9.3.

A pie chart can be effective in conveying a quick general impression (as in Figure 9.7). However, because they are difficult to differentiate visually, small differences in the size of sectors should not be represented in a pie chart – nor should sectors smaller than 7°. If the reader has to make accurate comparisons, a chart that uses lines to represent information should be preferred – because it is easier to compare line lengths than areas – but if the reader needs exact numbers, only a table will suffice.

10 Illustrations contribute to clarity

It is possible to communicate without words. In speaking we use gestures and facial expressions as well as words. In writing, the use of numbers enables us to be precise; and photographs, drawings and diagrams make possible the communication of information or ideas accurately, clearly, concisely, forcefully and quickly – either without words or with fewer words than would otherwise be needed (see Figure 8.1) .

Text tables and illustrations also help to break up pages of writing, provide variety and stimulate interest. By capturing the reader's attention, pictures help the writer to emphasise important points.

> And ye who wish to represent by words the form of man and all the aspects of his membranification, get away from the idea. For the more minutely you describe, the more you will confuse the mind of the reader and the more you will prevent him from a knowledge of the thing described. And so it is necessary to draw and describe.
>
> Leonardo da Vinci (1452–1519) *Notebooks*

Using illustrations as aids to explanation

Consider tables and illustrations as part of any document, not as ornament. They should complement your writing, so do not add them as an afterthought to a composition that is otherwise complete. Instead, before starting to write, consider how information or ideas can be best conveyed – to the readers you have in mind – in words, numbers, tables or illustrations.

Information conveyed in one way should not be repeated in another way in the same document (as it was in Chapter 9) to facilitate comparison of different methods of presentation: compare Table 9.3 with Figures 9.4b, 9.6 and 9.7; and compare Figures 9.6 and 9.7. Decide how best to convey the information, depending on your purpose and the needs of the reader, and then present it once only: in the text, in a table or in a figure.

By planning, you can avoid repetition and also ensure that in your composition each table and each illustration can be: (a) numbered; (b) arranged so that, if possible, it fits upright (portrait, not landscape) on the page; (c) placed near relevant text; and (d) referred to at least once in the text – with any necessary explanation, and with cross-references included in other (usually later) parts of the same composition if these will help the reader.

All illustrations (photographs, drawings, and diagrams) are called figures. In any document they should be numbered separately from the tables: *either* consecutively from first to last in a short document *or* using point numbering section by section in a long document (for example, in a technical report). Each figure should have a concise caption or legend – immediately below the figure – so that the figure can be understood without reference to the text.

A sequence of illustrations, like stills in a film strip or in a strip cartoon, may give a good indication of what the text is about. So anyone reading a well-illustrated article written in another language may look first at the illustrations, which provide an international language. If you wish to convey a message so that it can be understood by as many people as possible, you will prefer pictures to words.

Illustrations are an aid to precise description. Because they make an immediate visual impact, illustrations should not be included as a form of padding. They should provide information needed for an understanding of the work and reduce the number of words required in the text. If its message is simple and clear, a brief cross-reference to the figure may be all that is needed in the text.

Many people have an uncritical respect for the printed word and too readily accept what is written as necessarily true. The scientist learns to recognise differences of opinion and to read critically. However, illustrations have an immediate impact. No one reads without choosing to do so, but a glance at a picture may leave a lasting impression. This is why advertisers prefer pictures to words.

The value of illustrations cannot be over-emphasised but, because they are so effective in conveying a message, they must be planned and produced carefully so that the reader is not misled. The same tests of clarity and truth must be applied to your illustrations as to your writing.

Consider the reason for each illustration and what information you wish to convey before deciding what kinds of illustrations to use.

Photographs

Because they enable readers to see what is described in the text, photographs reduce the number of words required in the description. They enable readers to see things for themselves, and so serve the double function of depiction and corroboration. However, readers may be too easily convinced that what they see in a photograph is necessarily correct. A photograph cannot lie but it may mislead. This is especially likely if natural shadows, which give a three-dimensional effect, are destroyed by artificial lighting. Furthermore, when a report is published even the best prints lose something in reproduction (and they may also be half the dimensions of the original). As a result, some things you see in your original print may not be seen by the reader – who may not know that they are there.

A line drawing may be better than a photograph for illustrating microscopic or very small objects, because in a photograph parts of the subject may be out of focus whereas a microscope can be focused in different planes at different stages in the preparation of a drawing.

In selecting photographs for publication, therefore, look for relevance, scientific interest, sharpness of focus, and effective lighting and contrast, and then consider whether or not a good line drawing or diagram would serve your purpose better.

Line drawings

A drawing is not intended as proof but as illustration. The burden of proof rests on the scientist. In a drawing you can help to avoid confusion by directing emphasis to those things you consider essential to your argument (as in Figure 10.1).

Drawing is an aid to observation and in a line drawing each line should be an accurate record of an observation, so that the completed drawing is a summary of observations. A line drawing is so-called because lines are used to represent the things observed, without shading. Effective labelling is used to direct attention to parts named in the text or to emphasise those things you consider essential to your argument. If the proportions are to be correct the drawing must be to scale, and a scale should be marked on the drawing in metric units (as on a map or plan).

Inaccuracies in a drawing may result from lack of knowledge of the subject but this is usually corrected by the careful study required in the preparation of the drawing. Drawing is therefore a useful part of a scientist's training in the art of observation.

Inaccuracies may also result when a subject is well known to the observer – who represents what is believed to be true rather than what can

PARTS OF A FLOWER

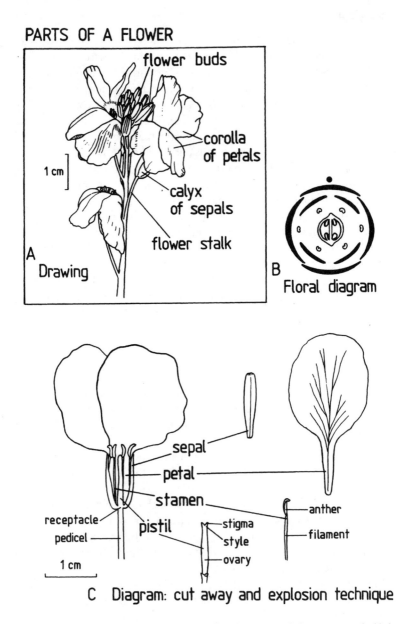

Figure 10.1 Drawing parts of a flower. The drawing and diagrams are half the dimensions of the original artwork. Lettering with 0.7 mm and 0.5 mm pens; letters 7 mm high (capitals) and a maximum height of 7 mm or 5 mm for lower case letters, prior to reduction. Other lines drawn with a mapping pen.

be observed. A student's drawings, for example, may include details from a text-book diagram that could not have been seen in the specimen being depicted. The experienced scientist should also be careful to draw only what is observable, not what is expected.

Drawings and photographs have several things in common. They represent three-dimensional objects in two dimensions – as they are seen at one time from one place. The photograph is objective but, like any other kind of illustration, it is interpreted by the viewer. Although both the photograph and the representational drawing may help readers, therefore, they may also mislead them. So a diagram may serve your purpose better.

See Chapter 9 for the use of diagrams in the presentation of numerical data and results.

Plans and maps

A plan or map, which must be drawn to scale and include a scale bar marked in metric units, conveys more information – more accurately – than would a photograph or drawing of the same subject.

A map is a kind of diagram, but there are difficulties in representing a globe to scale on a flat surface and some readers may be misled because they do not understand the projection used. On every map there should be an arrow indicating north, and north should be at the top of the page.

Any plans and maps (or other diagrams) that are to be compared should be drawn to the same scale and, if possible, they should be arranged side by side.

If symbols or different kinds of shading are used in any diagram, a key must be provided – preferably as part of the diagram (as in Figure 9.6) rather than in the legend, so that the symbols are not lost if a diagram is reproduced in another document with a different legend.

Diagrams that are not drawn to scale

Some diagrams are not drawn to scale. In these, each line is not intended as an accurate record of an observation: it is the diagram as a whole that provides a useful summary (Figures 4.1, 8.1 and 10.1B). It may help, for example, in presenting an idea (as in Figure 4.2), in making comparisons (as in Figure 9.2), or in showing the arrangement of equipment (as in an electrical circuit diagram).

Algorithms, also known as decision charts, are diagrams that help readers: (a) to make choices as they carry out an activity (for example, in a key for identifying objects), to follow sequences (for example, in a fault-finding procedure or at successive stages in a manufacturing process), or to understand relationships (for example, in a family tree); or (b) to appreciate links

(for example, in the chain of management in an organisation). Each algo-
rithm is essentially a set of instructions in which the user is given a choice
and has to make a decision at each step.

Block diagrams can be used to show the arrangement of parts in an item of
equipment.

Flow charts can be used to present the order of events in a process. Words
are used, and perhaps also drawings, and lines – with or without arrow
heads – to indicate the flow of, for example, materials, energy, or ideas.

Preparing illustrations

To match your writing to the reader's interests (see pages 85–6), and to
ensure accuracy and clarity, each illustration should be planned to go with the

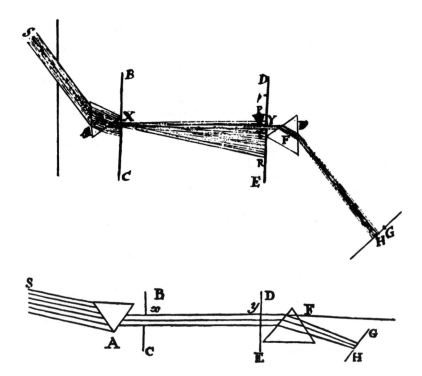

Figure 10.2 Above: Newton's diagram. Below: inaccurate redrawing by a later
author. Both diagrams are from Lohne, J. A. (1967) The increasing corruption of
Newton's diagrams, *History of Science*, 6, page 73.

text. Illustrations prepared for another purpose may save the writer time, but anything irrelevant may confuse the reader. However, if you do decide to use artwork prepared for another document, it is best to copy it photographically. Otherwise, lack of care or lack of understanding when redrawing illustrations prepared by others may result in the introduction and perpetuation of errors (see Figure 10.2). See also *Copyright*, page 144.

Scientists and engineers who are convinced of the value of illustrations, as an aid to communication, are likely to accept that time devoted to their preparation is well spent. An artist studies and interprets, and then draws what seems to be important. The drawing conveys the artist's understanding of the subject. Because of this, it is best if you can illustrate your own reports. You know what to include, what to omit, and what labelling is needed. At least, try to provide the artist with as good a sketch as you can manage.

In most documents the print and artwork are black on a white background. The use of colour, especially on a coloured background, can cause problems for some readers (for example, as a result of colour blindness, or because the colours chosen are indistinct when viewed in poor or coloured lights), and lettering on a coloured background may be lost if a document is reproduced in monochrome. Such considerations apply particularly to situations in which mistakes in reading may be prejudicial to safety (BSI, 1977), but clarity is important in all scientific and technical writing.

Publishers may prefer the artwork on disk, or as hard copies, or they may ask for both. If your illustrations are to be published, therefore, you must consult the publisher's house rules, or the instructions to authors issued by the editor of the journal to which you plan to submit your paper, before preparing any artwork – so that you are aware of any special requirements.

The advice given here applies to the preparation of original artwork, using pen and ink or computer software, with black print and black lines on a white background.

Dimensions

The dimensions of each figure should be chosen so that, if possible, it fits upright on the page (portrait, not landscape) and readers can look from the text to the illustrations without having to rotate the document. In a journal with a two-column format, an illustration may be the width of the column or that of the printed page (type area only).

When the size of a drawing has been decided, you should use a larger sheet of paper so that there are margins of about 40 mm. If the drawing is also to be used in preparing a slide for projection, its width will depend upon the width of a page or column in your report, but the proportions of the drawing for a 5 x 5 cm slide must be 3:4 or 4:3 (see also page 169).

If you are drawing with pens, your final drawings should be in waterproof black India drawing ink on photographically white Bristol board, graph paper with blue grid lines, blue tracing linen, or good quality tracing paper (110 g/m^2). The artwork should be prepared twice the dimensions required in the document (for photographic reduction by one half) with lines twice as thick (0.25 mm for the axes of a graph and, if any are needed, for the grid lines; and 0.5 mm, 1.0 mm and 1.5 mm for other lines on a graph).

All letters, numbers and symbols must also be twice the required size, with lines twice as thick (except that some publishers require all the letters and numbers to be in pencil). Lettering with capitals 4 mm high is large enough for most purposes. Letters should be spaced so that there is a clear, though narrow, gap between them; and the space between words should be the width of a lower case *n*. Lines of lettering should be well spaced. The space between ruled lines should be at least 4 mm. Every line in a drawing must be thick enough to show clearly after reduction, and any line that will not show should be erased. To facilitate the reading of numerical values, the gradations on the scales of a graph should be marked, for example, at intervals of 20 mm, so that after reduction by half they are 10 mm apart.

Preparing a large drawing (for photographic reduction) encourages bold work, with large pens, on a large sheet of paper, and helps in the inclusion of detail. Small imperfections of line are less obvious after reduction and a good drawing then looks even better. However, reduction will not make an untidy drawing look neat. Neatness of line is essential in the drawing if every part and relationship is to be clear after reduction. A neat and even appearance is obtained by working on the whole drawing rather than completing one part and then moving on to the next. Try to draw the whole of each line in one stroke of your pen. Draw straight lines with a ruler and curved lines with a compass, French curves or a flexible ruler. Use unbroken, broken and dashed lines, or different symbols, to distinguish different curves on a graph.

If you do the lettering in ink, use stencils and special pens for letters, numbers or symbols (or use transfers). Always draw labelling lines with a ruler. Preferably, they should be straight lines, radiating (as in Figures 4.1 and 10.1) so that they do not cross one another. Place your pen on the point to be labelled and draw a complete line, not a broken or dotted line, away from this point. Do not add an arrow head to a labelling line: if you do, the reader may not be able to tell if the arrow head ends on the part labelled or is pointing to another part; and arrows on a diagram may be used for other purposes (as in Figures 4.1 and 4.2). However, a publisher may ask for all words, letters, numbers and labelling lines to be in pencil and if arrow heads are unacceptable, you must emphasise this in your correspondence with an editor.

Graphs can be prepared on blue-lined graph paper and then photographed or traced. Although grid lines are essential in preparing a graph when using a pen and ink, they are not usually necessary for its interpretation (and the blue lines will not appear in the photograph). If any lines are required, therefore, these must be added in black ink.

If a graph is intended for publication, the symbols used for the points on a graph should be the symbols available to the printer (see the *Instructions to Authors* of the journal in which your work is to be published, or the publisher's house rules) so that, if necessary, identical symbols can be used in the legend, for example:

○　●　□　■　△　▲

If other symbols are used, a key should be included as part of the figure (not in the legend). Identical symbols and line forms should not be used on two curves in one graph if the points could be confused, but the same symbols should be used for the same quantities throughout a document.

Each axis of a graph should be labelled, parallel to the axis and on its outside. Numbers on the axes should also be outside the graph, but these should be upright next to small projecting bars (see Figures 9.2 and 9.4).

If possible, all the illustrations for one document should be drawn with the same line thicknesses, on pages identical in size (say, A4 = 210 x 297 mm), for reduction by the same amount; and when appropriate a number of drawings should be drawn to the same scale and placed together as one illustration. The parts of an illustration should be designated by upper or lower case letters, depending on house rules, not by numbers. Grouping illustrations reduces costs of production and results in uniform lines and letters.

Whether you prepare the artwork using computer graphic software or pen and ink, ensure that any graphs or other diagrams that are to be compared – even if they are not in one group – are drawn to the same scale.

Drawing

Consider how best to present information in an illustration so that you can convey information or ideas to the reader in your chosen way. Balance, which makes a drawing appeal to the eye, is achieved only if you consider how the drawing or the parts of a drawing, and the labelling, are to be arranged on the page. Compose each drawing so that information is conveyed effectively, and use labelling to help the reader. If the drawing has several parts, use letters or arrows to guide the reader.

So that your message is clear, do not clutter an illustration with too much information. If a graph or a drawing has too many lines, so that nothing stands out, the reader may have difficulty in distinguishing what is essential to your argument. Only you can decide what to leave out in the interests of clarity. An illustration should concentrate attention. For maximum impact, the drawing and the message must be clear and simple, and the most effective illustration conveys just one idea.

One way to prevent a drawing from becoming cluttered is to use two or more drawings instead of one. On a blackboard or whiteboard, information can be presented a little at a time as a diagram is constructed. An artist uses the same technique in a strip cartoon: the subject is presented simply in each drawing and its caption. This technique can also be used in giving instructions (as in many user manuals), and is taken further in the preparation of graphs – in which each point represents an observation or is a summary of observations.

Another way to prevent a drawing from becoming too cluttered is to use more of the page. Some subjects can be displayed effectively by using an explosion technique which helps to show how components fit together to form a more complex whole (see Figure 10.1C). The same kind of subject may lend itself to a cut-away technique – with superficial structures shown in their correct position but with enough cut away for the underlying parts to be seen (also in Figure 10.1C).

Completed illustrations

Check each figure and its legend, and ask someone else to check, to ensure that it serves its purpose. People will believe what they see. Subconsciously, a drawing presents difficulties for the artist and for the viewer (see Figure 10.3). The artist has to represent the subject in two dimensions and the viewer has to interpret the drawing so as to imagine the object in three dimensions. If possible, the drawings for a report should be simple, with clear lines, but additional artwork or labelling may be needed to facilitate interpretation.

State the reduction required, in soft black or blue pencil, on the margin of each drawing and draw a line around this instruction to indicate that the words are not to be included in the finished illustration. The instruction to reduce by one-half will give a final size one-half the dimensions of the original (a quarter of the original area).

Store hard copies of completed illustrations (photographs, pen and ink drawings and hardcopies of artwork prepared using computer graphic software) in an envelope, unfolded, with a sheet of cardboard to prevent bending, so that they do not become soiled or creased.

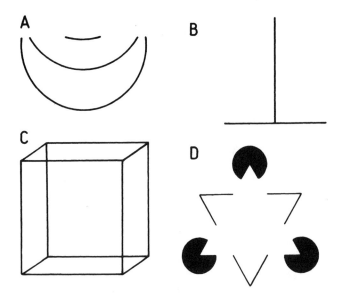

Figure 10.3 Optical illusions: (A) arcs of equal radius; (B) horizontal and vertical lines of equal length; (C) a cube which may be interpreted differently at different times; and (D) an illusion of a whiter triangle.

Improve your writing

Writing legends to figures (captions)

Each figure must have a legend as well as a number. Because more people look at the figures than read the text of a document, it should be possible to understand each illustration without reference to the text. The legend should therefore be complete, clear and concise.

A statement in the legend should indicate whether the points marked by symbols on a graph are records of observations (data) or arithmetic means (results). If vertical lines are drawn through the symbol, above and below the mean, the legend must indicate whether these show the standard error ($S\bar{x}$) the 5 per cent fiducial limits of error = 1.96 $S\bar{x}$) or the range.

An illustration should not be cluttered with information that could be put in the legend, but a scale on a drawing is better that a statement of magnification in the legend – because if artwork is photographically reduced, for publication, the scale remains correct. Similarly, a key to any shading or symbols should be included on the figure rather than in the legend (so that if the figure is reproduced with a different legend it can still be understood).

When an illustration is used to inform, it must be a correct record of an observation or an accurate summary of observations; and the legend (in the present tense) must be a factual explanation (as in Figure 9.2). But when a diagram is used to convey ideas, this must be made clear in the legend (as in Figure 4.2).

Any help, the source of data, and the source of any illustration that is not original, should be acknowledged (as in the legends to Figures 9.2 and 9.3). See also *Copyright*, page 144.

Checking your illustrations and legends

1 Check every drawing or diagram against your original artwork.
2 Check for clarity, accuracy and neatness of line.
3 Does the numbering of the figures correspond with the numbers in the text, and are other things referred to in the text included in the artwork?
4 Are the letters, words and abbreviations on the figures consistent with those used in the text?
5 Is the labelling clear and are the labelling lines acceptable? See page 112.
6 Is any figure cluttered with too much information?
7 If the figure is drawn to scale, is the scale marked on the illustration in metric units?
8 Are any diagrams or drawings that are to be compared arranged side by side and drawn to the same scale?
9 Are numbers and symbols for SI units of measurement marked clearly on all axes and scales?
10 Are all symbols sufficiently explained?
11 Should any photograph be replaced by a drawing?
12 If the illustration is to be published, check that the information required by the printer is written in the margin or lightly on the reverse in soft black or blue pencil (author's name, title of document, number of figure, reduction required).
13 Check that the information in a table is not duplicated in an illustration.
14 Check that each table and illustration will fit upright on the page (portrait) in the space available, and check that there will be space below the figure (that is, on the same page) for the legend.

11 Finding information

Reading as part of science

We learn many things by observation but most of what we know comes from conversation or reading. Discoveries are made against a background of existing knowledge which forms part of the opportunities of place and time.

Reading may save you the fruitless labour of seeking, by observation and experiment, information that is already in the literature; but do not be convinced too easily that something you wish to investigate has already been studied exhaustively. What is written is not necessarily true and is seldom the whole truth. Even accurate observations may be incomplete and changes in technique or in the design of experiments may lead you to new observations, to a different interpretation, and to new lines of enquiry.

Some people have too much respect for the printed word, but experienced scientists may agree that: 'It is that which we do know which is the greatest hindrance to our learning, not that which we do not know' (Claude Bernard). 'The advantage of a certain amount of ignorance is that it keeps you from knowing why what you have just observed could not have happened' (Sir Frederick Gowland Hopkins). Reading before you investigate may direct your mind along well-worn tracks and away from a fresh approach to a problem.

In an essay *On the Ignorance of the Learned* William Hazlitt (1778–1830) wrote that:

> Books are less often made use of as 'spectacles' to look at nature with, than as blinds to keep out its strong light and shifting scenery from weak eyes and indolent dispositions. The bookworm wraps himself up in his web of verbal generalities, and sees only the glimmering shadows of things reflected from the minds of others.

Relate the time you spend on your literature search (see Figure 11.1) to the amount of time available for your investigation – including your composition. It is better to work in ignorance of what has been done than to spend so long in searching the literature that you have insufficient time for your own observations. Also, be prepared to observe and experiment as the opportunity arises. Your first observations will make possible a more informed reading of other people's work and serve, if nothing more, as a trial run.

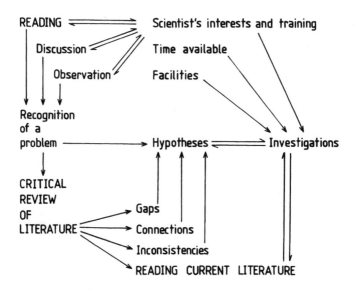

Figure 11.1 Reading as part of a scientific investigation.

Other people's work

Observations by other people may complement your own or suggest new ways of looking at a problem and new lines of investigation. You may also link ideas previously unconnected in the minds of others, or improve your own procedure, after considering how others have tackled similar problems.

Relating new observations to earlier work should also lead you to a deeper understanding of your problem. Although you may work alone, most discoveries are the result not only of observation and interpretation, but also of communication between specialists.

With so many scientists, and more journals published every year (see Figure 11.2), it is difficult to know what other people have written on any subject. Starting with recent publications, you will find references to related papers – and it is unwise to restrict your reading to those aspects of a problem which are of immediate interest, because new ideas may come from unexpected sources.

Sources of information

Information technology is concerned with electronic methods of cataloguing, communicating, processing, storing, retrieving, and publishing information. People speak of the electronic office as a place where there is no need for paper, but much information is still recorded, stored and communicated on paper.

Dictionaries

Dictionaries are available for most languages and for most other subjects. For anyone writing at work, a good dictionary of the English language is an essential reference book. It provides a guide to much more than correct spelling (see page 65), so the spell-check on a computer is not an alternative.

For anyone who needs more information than can be included in a desk dictionary, the *Oxford English Dictionary* is a printed multi-volume work with CD-ROM and online versions that provide access from a computer terminal to a database comprising more than 500 000 words.

Encyclopaedias

An encyclopaedia, which may be available in a library as one or more volumes or in electronic form via a computer terminal, is a good starting point for anyone coming new to a subject. Each article is written by an acknowledged authority, in language that can be understood by non-specialists, and it ends with references to other sources of information for those who need to know more. Multimedia publications provide spoken words and other sounds as well as printed text, and moving pictures as well as stills. In addition to such general works there are specialist encyclopaedias on many subjects.

In engineering and the sciences well-known encyclopaedias include the McGraw Hill *Concise Encyclopaedia of Science and Technology*, van Nostrand's *Scientific Encyclopaedia* in two volumes, and the McGraw Hill *Multimedia Encyclopaedia of Science and Technology*.

Handbooks

The Complete Plain Words Gowers (1986) is a handbook for all those who use words as tools of their trade, and *Usage and Abusage* Partridge (1965), a book on English usage – published first in the United States – includes notes on American usage (see also Fowler, 1974).

There are concise reference books, for day-to-day use, on most subjects. For example, the CRC *Handbook of Chemistry and Physics* is a well-known source of data in the sciences. Other handbooks, usually called technical manuals, are supplied with many commercial products. Each manual describes a product and provides instructions, as appropriate, on how to store, handle, install, use, maintain and service the product correctly and, when the time comes, on how to dispose of it safely.

Standards

Many national and international organisations produce standards (see Table 11.1) to encourage uniformity in, for example, the use of units of measurement (Table 9.2) and the content, layout, preparation and management of documents (Table 12.2). Many organisations work to particular standards and require their suppliers to produce goods or provide services conforming to these standards. However, as standards are up-dated from time to time it is essential that an organisation, its suppliers, and the organisations it supplies are all working to agreed specifications.

Directories

There are directories covering many subjects – including companies, trades and other organisations. Names and addresses may be included, as in a telephone directory, and other information. Many directories are available in printed and electronic versions. For example, all the names and telephone numbers in a complete set of the United Kingdom *Phone Book* are also available on line (and, as stand alone or multi-user versions, on CD-ROM). Other useful directories include two lists of publishers and of books in print: *Books in Print,* published in New York, and *Whitaker's Books in Print,* published in London (available in printed, microfiche and electronic versions). Some directories, available only in electronic form, are called listings.

Access to other sources of constantly up-dated computer-stored information is also available by television in the home or office, and via the Internet.

Books

It is not possible to keep all the books on one subject together on the shelves of a library. To find out which books on any subject are stocked by

Table 11.1 Some American (ANSI), British (BS), European (EN) and
International (ISO) standards concerned with aspects of writing,
available in print and CD-ROM versions (and also via the Internet),
listed in alphabetical order by subject

Abbreviation of title words and titles of publications: BS 4148 (identical with
ISO 4)
Abbreviations for use on drawings and in text: ANSI/ASME Y14.38
Abstracts: Guidelines for writing: ANSI/NISO Z39.14
Alphabetical arrangement (and the filing order of numbers and symbols): BS 1749
Bibliographic references: BS 1629 (similar to ISO 690; more detailed than BS 5605)
Citing and referencing published material: BS 5605 (a concise introduction)
Copy preparation and proof correction, marks for: BS 5261C (and, for
mathematical copy, BS 5261–3)
Indexes: content, organisation and presentation: BS ISO 999 (also ANSI/NISO
TR-02)
Indexes: selection of indexing terms: BS 6529 (similar to ISO 5963)
Information technology: information security management: BS ISO/IEC 17799
International System of Units (SI units): BS 5555 (identical with ISO 1000)
Numbering divisions and sub-classes of written documents: BS 5848
Occupational health and safety management systems: BS 8800
Presentation of research and development reports: BS 4811
Proof correction, marks for, and copy preparation: ANSI Z39.22 and BS 5261C
(and, for mathematical copy, see BS 5261–3)
Quality management and quality assurance: BS ISO 9000 and BS EN ISO 9004
Quality systems: BS 5750
References to published materials (including bibliographic and cartographic
materials, computer software and databases): BS 1629 (similar to ISO 690)
Scientific and technical reports: elements, organisation and design: ANSI/NISO
Z39.18
Scientific papers for written and oral presentation, preparation of: ANSI Z39.16
SI units: BS 5555 (identical with ISO 1000)
Specifications, guide to the preparation of: BS 7373
Statistics, vocabulary and symbols: BS ISO 3534
Technical manuals: guide to content and presentation: BS 4884
Typescript copy preparation, for printing: BS 5261–1

---

a library, first look at the subject index for the classification number for that
subject. Then look up this number in the subject catalogue, where you
should find an entry for each book stocked. The book number in each entry
indicates where the book with this number is to be found on the shelves.

If you know which book you require, use the alphabetical catalogue, in
which the names of authors or editors (and those of organisations, government departments and societies that produce books) are listed in alphabetical
order. Each entry in this catalogue includes bibliographic details of a book (or
of another source of information) and its classification number.

In a small library each entry in the classified and alphabetical catalogues may be on a separate index card, but in most libraries access to the catalogues is via a computer keyboard. Detailed instructions on how to use the catalogues are displayed on the computer screen. You will be able to search: (1) by entering a classification number to see what books the library stocks on a particular subject; or (2) by entering the name of an author, the name of an organisation, or the title of a book, if you are looking for a particular book; or (3) by entering a key word that you think is likely to be included in the title of a book on the subject that is of interest to you. By entering a key word you may also find details of relevant non-book materials available in the library (for example, maps, collections of photographs, audio and video tapes, and public records on microfilm).

Reviews

Some books and journals specialise in the publication of articles reviewing the literature on a particular aspect of science, and some reviews are published in journals that also publish original papers. In a review all relevant published work should be considered, so a review is a good starting point in a literature survey. However, reviews may say nothing of the methods used in the work reviewed and each reference to previous work is necessarily brief and may be misleading. Books and reviews are called secondary sources and it is important to look at original articles (primary sources) to be sure that in referring to the work of other writers you do not misrepresent them.

Specialist journals

The results of original research are published in specialist journals. In these primary sources you can read the results of recent work soon after it is published, and see references to related papers that may be of interest.

However, it is not possible for a library to subscribe to all the journals you could find of interest (see Figure 11.2); and may not be published in the most appropriate journals. So you will not find all the papers you might like to read in the journals stocked by your local libraries. To see titles, complete bibliographic details and abstracts of papers published in both current issues and back numbers of many journals you will need to use computer-based information retrieval systems (see Table 11.2). A search for articles on a particular subject can be based on key words (words that you would expect to be included in the titles of articles or in journal indices). Many journals are published in electronic as well as print versions, and some only in electronic versions, and these are available via the Internet.

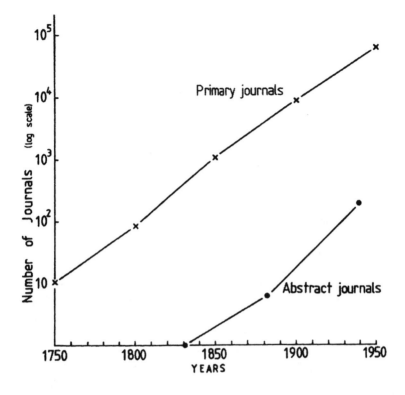

Figure 11.2 Number of journals founded 1750 to 1950, including those that are no longer published. Logarithmic scale on y axis.
Estimates from de Sola Price, D. (1975) *Science Since Babylon*, Yale University Press..

Abstracting journals reprint abstracts from published papers, or edited abstracts, with a full bibliographic reference to each paper. This information will help you to trace relevant papers, but abstracts lack detail and, like reviews and books, they should be regarded as an introduction to the literature, not as a substitute.

Indices are published as part of many journals, but some journals have no index and some have an index that is incomplete and therefore misleading. However, most journal articles include key words for use by computerised indices that cover many journals (see Table 11.2). To begin a search you may enter an author's surname or a search term (a key word or phrase).

A literature search based on the *Science Citation Index* can start with a reference to a paper or with the author's name, and this leads to other authors who have cited the paper and so to related literature. From this index scientists can find out whether or not their work has, for example, been applied, extended or criticised. The Royal Society's *Catalogue of Scientific Papers* is an author index to the papers published in 1500 journals in the nineteenth century.

The Internet (World Wide Web)

With a web browser, you can use the Web address of a business or other organisation to access its web site from a personal computer and see the pages it provides – which include, for example, words, pictures, videos, plans and maps. Via the Internet, therefore, much useful information is available – but also much unsupported opinion, and much that is fiction.

Like the articles published in newspapers and in trade magazines, and unlike the papers published in professional journals (in printed or electronic versions), much of the material on Web pages has not been subject to peer review and some is no more than advertising. Also, the contents of Web pages may be changed at any time, so it may not be possible to state the source of information obtained from the Internet in such a way that readers can consult the same source and read an identical document themselves. Keep these reservations in mind when you use the Internet. Also, because Web pages may change at any time, if a document is of particular interest you are advised to download it to your computer or to make a hard copy.

Many organisations include a Web address on their headed notepaper and in advertisements, and there are directories of Web addresses, but if you do not know an address, you can try to guess it – because most Web addresses comprise: www (the World Wide Web), the name of the organisation (for example, ons = Office of National Statistics), an extension indicating the type of organisation (for example, co = company; gov = government), and the country (for example, uk = United Kingdom) – with full stops where there are commas in this sentence, but with no spaces. For example, www.ons.gov.uk is the Web address of the Office of National Statistics, a government department in the United Kingdom. However, when you access a Web address you must check that the site is that of the organisation you are seeking – because different organisations, perhaps with opposing objectives, may have very similar addresses.

Via the Internet you can also, for example: (a) study previously inaccessible archives, (b) browse through the catalogues of major libraries, (c) scan pages of both current issues and back numbers of newspapers, (d) search indices for bibliographic details and abstracts of publications likely to be of interest to you, and (e) read (and, if necessary, print out or down-load to your computer) articles from journals published electronically (see Table 11.2).

Table 11.2 Some electronic sources of information on articles in journals

Electronic sources[a]	Access to[b]
Analytical Abstracts	Papers in analytical chemistry and chemistry journals
BIOSIS	Bibliographic details and abstracts of books, papers on biology in journals, and conference proceedings
Chemical Engineering & Biotechnology Abstracts	Bibliographic details and abstracts of papers in biotechnology and chemical engineering journals
Chemistry Citation Index	Papers cited in chemistry journals
CINAHL	Cumulative Index to Nursing and Health Sciences
Computer database	Articles in computing journals
E4data Engineering	Articles from catalogues of engineering firms
Geobase	Papers on earth sciences, ecology, geology, geography and marine sciences
INSPEC	Computing and IT journals, and conference proceedings
MEDLINE	Medicine
PsycINFO	Articles on psychology and related disciplines
SCIENCE DIRECT	Articles in science and engineering journals
SERLINE	Articles on biomedical and health sciences
Web of Science	Science Citation Index, Social Science Citation Index, and Arts and Humanities Citation Index
ZETOC	Electronic Table of Contents of journals in the arts, business, engineering, finance, humanities, law, sciences, technology and conference papers

Notes
a As with other businesses and organisations, the names, ownership and location of electronic sources may change.
b Libraries with access to computerised indices provide notes to help users.

Many individuals have Internet accounts with an Internet Service Provider (ISP), and pay for this either by a direct charge or through a telephone company. So use of the Internet from a personal computer can be expensive. A search for information can take a long time and you pay both directly because of the cost of the service and indirectly because of the value of your time. Also: (a) your search will not necessarily be successful, and (b) you may not be able to rely on the relevant material you do find – much of which is likely to be opinion, unsupported by evidence. As when reading review articles (secondary sources), you will need to refer to primary sources (see page 122) for the evidence upon which statements are based.

Search engines are used in looking for information on the Internet. Many of these offer both a simple search and a more complex search that may be called an advanced search. However, no search engine could search the whole of the Internet, and if you enter identical search requests into different search engines you will find differences in their outputs, even when searching for specialist terms. One reason for these differences is that organisations developing Web pages use many different key words, not just the most appropriate words, in an attempt to direct searches to their pages. Another reason is that some search engines accept new Web pages quicker than others, and some store pages for longer than others.

Intranets

An Intranet is a Web, similar to the Internet, but with restricted access. For example, it may be available within an organisation – linking computers on the same site or on different sites, or even with an international company linking computers on sites in different parts of the world. Because access is restricted the information displayed is easier to control and is likely to be of better quality than much of the information available on the Internet. If you are working for an organisation that has an Intranet, this should be where you concentrate your first searches.

Improve your writing

Reading to some purpose

When trying to find answers to your questions, or seeking background information, you do not need to read the whole of every book you consult. Some books are written for reference, but even those that can be read as a whole can also be read in part – to find just the information you require at the time. This is a good way to start reading about a subject, because you will remember best those things that you want to know or find most interesting.

It is best to start with recent publications on any subject to find the present position and to be guided by your special interests to earlier literature. In deciding how far back to go, an important consideration is the amount of time available. Most people enjoy reading the work of others with interests similar to their own; and observations recorded, however long ago, are as correct now as on the day they were written.

When deciding what to read – be it the title, the summary, the introduction, one sentence, one paragraph, or the whole text – remember that an effort is required by the reader as well as by the writer. Scientists should practise an economy of expression, and should write carefully so that readers can grasp their meaning quickly; but this can be so only if each word is read.

Read carefully to make sure that you take the intended meaning. Read critically, as a stimulus to thinking. Weigh the words and consider the evidence and arguments. Such questions as these should be in your mind: What is being said? Are the underlying assumptions correct? Are the arguments balanced (unbiased)? Are the statements supported by sufficient evidence? What are the implications of the work? Is there a better explanation of the evidence presented?

Reading the prose of other scientists is the student's introduction to the conventions of scientific writing. In reading, scientists learn more about science and, from those who write well, they learn how information and ideas can be presented clearly and concisely and in an interesting way.

Making notes as you read

Many readers spend time making detailed notes. If you do this as a habit, consider whether or not your time could be better spent. Students should buy up-to-date books on each aspect of their courses. They should read these in preparation for course work, use them to augment and clarify their lecture notes, and learn from them by reading selected passages several times.

Keep a note of everything you read. Use a sheet of A4 paper so that you can file it with your other notes on the subject, or if you are undertaking a literature search you may prefer to use a separate index card for each publication and to keep the cards in alphabetical order. If the card is large enough (203 × 126 mm), there is space for a summary, for notes or quotations, or for a reference to more detailed notes, a photocopy or a reprint kept in another place, or for the shelf number of a book in a local library.

First in each note, as a heading, record the author's name and other bibliographic details of the publication. You will need these details so that: (1) you can remember the source of your notes; (2) you can refer to the same source again at any time; (3) you have all the information to hand if you decide to include details of this source in a list of references at the end of any document you are writing; and (4) you can obtain any other publication cited in this source. Also record the number of each page from which you make any further notes, so that you can find the page again, if necessary, or refer to it in citing the source of your information. You may also find it helpful, when making notes from a book, to record its unique reference number (the international standard book number: ISBN) written on the reverse of its title page.

In the bibliographic details of a chapter in a book, the name of the writer comes first, then the date of publication of the book, then the title of the chapter, which should not be underlined or printed in italics, followed by the word *in* (underlined in typescript or printed in italics) and this by a colon, the name(s) of the editor(s), the abbreviation ed. or eds., the other details as for a book, and then the first and last pages of the chapter.

In the bibliographic details of an article in a journal, magazine or newspaper the author's surname and initials are followed by the date of publication in parenthesis, the title of the article using capital letters only for proper nouns, the name of the publication (in manuscript underlined, and in typescript either underlined or in italics), the volume number (underlined with a wavy line in manuscript and in typescript printed in bold) without the abbreviation vol., the issue number in parenthesis, and the first and last page numbers joined by a dash. For example:

> Barrass, R. (1990) Scientific writing for publication: a guide for beginners. *Journal of Biological Education*, **24** (3) 177–181.

In the bibliographical details of a book the author's or editor's name and initials (or the name of the issuing organisation if no author or editor is named) may be followed immediately by the year of publication in parenthesis (or, depending on house rules, the date may be placed later), the book's title (underlined in manuscript and printed in italics, with capital initial letters for most words), the edition number (except for the first edition), the number of volumes (for example, 2 vols) or the volume number (in arabic numerals and underlined with a wavy line in manuscript or printed in bold) without the abbreviation vol., the place of publication, the name of the publisher and the date of publication if this has not already been included, and either the number(s) of the page(s) referred to or the number of pages in the book including the preliminary pages. For examples, see *References*, page 195.

When recording the source of information from an Internet site, note the name of the originator (author, editor or organisation), the date and title, followed by the word [*online*] in square brackets, the place of publication, the publisher (if known), the word *Available*, and then the name of the service provider, an Internet address, and [the date accessed] in square brackets. For example:

> Shields, G. and Walton, G. (1998) *Cite them right! How to organise bibliographical references* [online], Available from HTTP: http://www.unn.ac.uk/central/isd/cite/ [22 Feb 01].

Note: the home page of this document is http://www.unn.ac.uk

When recording bibliographic details of a book or of an article published in a journal which is also available on the Internet, include the usual reference details as for a book or article followed by the medium (for example, online) and then by details of the Internet site. For example:

> Smith, A. (1997) *Publishing on the Internet*, London: Routledge. Online. Available HTTP:
> http://www.ingress.com~astanart.pritzker/pritzker.html
> [4 June 1997].

When preparing notes, as a student or in any other employment, decide what information you require and identify relevant passages in books and papers. Then relate the information and ideas from your reading to your own knowledge of the subject. Your notes will usually be brief (key words and phrases, headings and sub-headings, concise summaries and simple diagrams) but take care that they are accurate. Such brief notes are useful as memoranda, and as an aid to study, and will be useful in filling gaps in your topic outline for an essay or report. Do not waste time copying long passages word for word (for example, extracts from your textbooks that, if necessary, you can read again and again).

In summarising make a clear distinction between the author's conclusions and your own comments, so that you do not misrepresent the author's views later. If you think you may quote anything in your own work it is best to take a photocopy, but if this is impracticable, make sure you write every word and punctuation mark exactly as on the page from which you are copying, plus quotation marks to remind you that it is a quotation. To copy complete sentences from a composition written by someone else and present them as your own, that is to say without acknowledging their source, is plagiarism (stealing thoughts) and is unacceptable; and to copy even a short extract from a book or other publication without permission, even if the source is acknowledged, may, if your work is to be published, be an infringement of copyright (see page 144).

Citing sources of information

Do not cite a source of information, in any report or other document you write, unless you have read it – as a whole or in part – to check that you are not misrepresenting the author.

If you use an author's exact words, the words quoted should normally be indicated by quotation marks (as on page 13), but a longer quotation may be indicated, without quotation marks, by indenting the quoted material (as on page 48). However short the quotation, the author's words and punctuation marks must be copied carefully, and you must acknowledge the source by giving the title of the publication, the date of publication and the author's name, either next to the quoted material or in a list of sources at the end of your composition.

If you summarise information or ideas from compositions written by other people, instead of using their exact words, you should still acknowledge your sources: (a) to acknowledge the work of others; (b) so that readers know that the views expressed are not necessarily your own and (c) so that readers can, if they wish, look at the same publications themselves. There are two widely accepted ways of citing sources in reports and similar documents.

One way, the numeric system, is by adding a number after the author's surname or at the end of a statement: in parenthesis (1) or [1], or in superscript[1]. A numbered list of sources is then included at the end of the composition (in the order in which the sources are first cited in the composition).

The other way, the name and date system (also known as the Harvard system), is by writing the name of the originator of a publication (usually the author's name), followed (in parenthesis) by the year of publication. The names of authors are then listed in alphabetical order, and works by each author in date order, at the end of the composition. If more than one publication by an author, in the same year, is cited in a composition the one cited first is marked by a lower case letter *a* immediately after the date, and the next by a lower case letter *b*, and so on.

The name and date system is used in most scientific and technical publications: it has the advantage, for most specialists, of being immediately informative. But it is unsuitable for use by authors dealing with old books or papers that include no publication date.

When citing a source using the name and date system, you may write the author's surname (followed by the year of publication in parenthesis) and then write what the author considers or states. For example: Quiller-Couch (1916) listed words that should be used with care by writers who wish to avoid jargon. Alternatively, you may include a summary of the author's views, findings or conclusions, and then end your sentence with both the author's surname and the year of publication in parenthesis. For example: Words that should be used sparingly and with care, by those who wish to avoid jargon, include case, character and nature (Quiller-Couch, 1916).

You may also be required, by house rules, to include the relevant page number or numbers immediately after the date, particularly if you quote the author's actual words. For example, you could refer to Quiller-Couch (1916: 87) as having listed words to avoid.

Writing a book review

Some working scientists review books, and writing a book review is also a useful exercise in comprehension and criticism that can be undertaken by students in their specialist studies or as part of a writing course. In preparing a book review, reading and writing should be associated closely (see pages 78–9).

12 How to write a report on an investigation

Planning your report

Before starting work on a report you must know why it is needed, and have clear instructions or terms of reference, stating exactly what the report is to be about and setting limits to its scope.

Analyse your audience. Identify your readers as clearly as possible. For example, if it is to be an internal report, prepare a distribution list to focus your attention on the needs of the different readers.

Allocate your time. You must know when the report is required so that you can decide how much time can be devoted to each of the four stages in composition: to *thinking*, to *planning* and collecting information, to *writing*, and to *checking* and if necessary revising your work. A deadline also helps you to decide on the depth of treatment that should be achievable. Even when working only for yourself, you must consider what you need to do and then allocate your time. Effective time management involves working to a timetable so that you can meet deadlines imposed by others or by yourself.

Preparing a topic outline

Make concise notes as you think of topics that may be included in your report. As an aid to thinking, try to anticipate questions that will be in the minds of your readers (see *Thinking* and *Planning*, pages 40–4, and page 133). Readers will expect relevant information, well organised, and clearly presented – with enough explanation.

Designing your message. Consider also why you are writing the report. For example, is it to provide information, to explain a decision already taken, or to persuade readers to accept your recommendations? Your intentions will affect both what you say and how you say it.

The sections of a report may be given different names in the house rules of different organisations, and in a short report it may not be necessary to use headings, but you must always answer these questions:

Answer the question: Why?
Introduction
    Why did you do the work?
    What is the problem?
    How did you become aware of it?
    Why is it of interest to readers of this report?

Answer the question: How?
Materials and methods
    How did you obtain the information, related to this problem,
    include in this report?

Answer the question: What?
Results
    What did you find?
Discussion
    What do you make of your results?
    How do your findings relate to previous work?
Conclusions
    What do you conclude?
Summary
    What, in a few short sentences, are the main points of your report?

Answer the question: Who?
Acknowledgements
    Who contributed ideas, information, illustrations?
    Who financed the work?
References
    How can I obtain a copy of each of the sources cited in this report?

Communicating your purpose. The use of widely accepted section headings in a published article or paper, or internal report of an enquiry or investigation (for example, see Table 12.1), if these are appropriate, will help you to plan your work: (a) to know where information on each aspect of your work should be placed; (b) to ensure an orderly presentation of material; (c) to ensure that nothing essential is omitted; and (d) to avoid unintentional repetition.

However, some repetition is needed in a long report, as in a book that may be used for reference, to ensure that those who do not want to read every word will be able to find the information they need in the parts they might be expected to read.

Other headings may be used, if more appropriate, except that: (a) all students on a course may be asked to set out their reports in a uniform way; (b) working scientists and engineers may have to follow their employers' house rules for the preparation of internal reports; and (c) anyone preparing work for publication must satisfy the requirements of the editor (as set out in the publishers' notes for guidance). In such notes and rules, authors may be asked to use certain headings unless there is some very good reason for doing otherwise.

In a further attempt to encourage uniformity of presentation, nationally and internationally, standards have been prepared by a number of organisations. For example, BS 4811 on the presentation of research and development reports, and ANSI NISO Z39.18 on the organisation and design of scientific and technical reports (see Table 11.1).

Obtaining a response.   As you consider the purpose and scope of your report, list relevant facts and ideas below appropriate headings and sub-headings as you decide: What should each paragraph be about? What needs most emphasis? What can be left out? Everything you include should help you to achieve your purpose. It must be: (a) relevant; and (b) necessary.

By numbering the topics for paragraphs in order below each heading, you can make your list into a topic outline. You will be reminded of relevant topics as you work on the outline, and recognise gaps in your knowledge that must be filled – perhaps before starting to write. If, when you do sit down to write, you have difficulty in getting started (see Figure 12.1), it is probably because you have not yet prepared a sufficiently detailed topic outline. That is to say, you have not yet decided exactly what you must say (content), or how best to say it (order) so as to capture and hold the readers' interest (relevance).

If you have a co-author, to minimise the risk of omissions and to avoid duplication of effort, you should agree first on the headings to be used and who is to write each section. Later, before starting to write, you should agree on a detailed topic outline. Also, when you are ready to start writing, whether or not you have a co-author, it is good practice to discuss your outline with the person who commissioned the report, so that you can check that your interpretation of the terms of reference is acceptable and benefit from any comments or advice.

First, have something to say!

Figure 12.1 The report as part of the investigation. If you think of your report as part of your investigation, not as a duty to be undertaken when the work is otherwise complete, questions such as when to start writing or which part to write first do not arise.

Numbering the sections of your report

In technical writing, especially, to facilitate cross-referencing, the parts of a long document can be identified by decimal numbering (point numbering) in both the text and the list of *Contents*. In using this method, no headings are centred. The first section heading is numbered 1. The first sub-heading in this section is numbered 1.1 and the next 1.2, etc., and minor headings below the first sub-heading 1.1 are numbered 1.1.1, 1.1.2, etc. It is possible to continue this decimal numbering (numbering each paragraph below each minor heading), but this soon becomes cumbersome. So if decimal numbering is used it should not normally go beyond two points.

An alternative to the decimal numbering of paragraphs is to signpost them by letters: (a), (b), etc., but if used with the decimal numbering of section headings this can be confusing to readers. So it is probably best to keep small letters for successive items in lists, and if it is necessary to number the paragraphs to number them consecutively throughout, and not to number the headings.

However, in a scientific article or paper (as distinct from a technical report) it is not usual to number either headings or paragraphs. Instead, in a hierarchy of headings, main headings could be in capitals and centred, second-order headings in capitals but not centred, and third-order headings with an initial capital letter for the first word or for most words (and all headings with a line to themselves). For most purposes three grades of heading are enough, but if fourth-order headings are required they can be underlined, and the text run on (after a full stop) on the same line. With a word processor, main headings could be in capitals and centred, second-order headings in capitals but not centred, third-order headings in bold, and fourth-order headings in italics.

The parts of a research report

If you do not have house rules, or a document prepared previously for your employer, as a guide to an acceptable format, the following notes should help you to ensure that your report is well presented. Use your judgement in deciding which sections are appropriate, and what section headings to use, in your report (see for example, Table 12.1).

The Front Cover

Include some or all of the following information on the Front Cover, as appropriate:

1   Top left: the name of the organisation (and of the division of the organisation) responsible for producing the report, and its full postal address.
2   Top right: The date of issue or the date when completed and ready for reproduction, as appropriate; and an alphanumeric reference of less than 33 characters, unique within the organisation, which identifies the report and the organisation. This alphanumeric reference should be repeated on the top right-hand corner of every page.
3   For the title, which should be about one-third of the way down the page so that it catches the eye, a *sans serif* font may be used and a larger print size than for the rest of the report.
4   The name(s) of the author(s) follow in alphabetical order, or in an order that reflects each person's contribution, or in an order determined by house rules or by national custom.

Depending upon house rules, the Front Cover (see Figure 12.2) may also

Name of organisation commissioning report   Alphanumeric reference

Date

EFFECTIVE POSITION FOR TITLE

(about one third of way down from top of page is eye-catching)

Author's name

Position in organisation

Summary could be placed here.

Any other information, as required by house rules (for example, a distribution list, a security classification), is best placed near foot of page where it does not detract attention from the title.

Figure 12.2 Layout of the Front Cover of a report: note that for the title a *sans serif* font is used, and a larger print size than for the text of the report.

Table 12.1 Arrangement of a research and development report +

Front cover
Title page *or* Report documentation page (ANSI)
Summary (abstract)
Preface (not usually needed)
Table of Contents (needed for all except short reports)
Introduction
Theory (additional to or alternative to next section)
Procedure and results (with sub-headings)
Discussion
*Conclusions (must be precise, orderly, clear and concise)
*Recommendations (arising directly from the conclusions)
Acknowledgements
List of references
Appendices
Tables
Illustrations (if not included in the main body of the text)
Graphs
Literature survey (if needed)
Bibliography (supplementary to list of references cited in text)
Glossary (if needed)
List of abbreviations, signs and symbols, if needed (*or* after Table of Contents, ANSI)
Index (if needed)
Distribution list (if required by the sponsor or by house rules)
Document control sheet (containing numbered boxes for such things as the report's reference number, the contact number, and the security classification).
Back cover

Notes
+ Consistent with standards BS 4811 for research and development reports, and ANSI NISO Z39.18 for Scientific and technical reports.
* Alternatively, the conclusions and recommendations may be placed immediately after the introduction.

include a summary, a distribution list (usually in alphabetical order), the security classification or a statement relating to confidentiality, the price, and the sales point if different from the organisation responsible for the report. However, try to ensure that the Front Cover is sensibly arranged so that any other necessary information does not detract attention from the title.

Any special notices required by a sponsoring organisation should be on the inside of the Front Cover. In a bound report, after the Front Cover, the first sheet is blank, and the next (the Half Title) has only the title. The next page is the Title Page.

The Title Page

The title comes first, followed immediately by the sub-title and then by the name(s) of the author(s). The *Abstract* or *Summary* may also be included on the Title Page, or it may be on the next page immediately before the *Introduction*. Clearly, there is some duplication of information on the Front Cover and Title Page. You do not necessarily need both; and some employers prefer to start, instead, with a report documentation page (see Table 12.1).

Just as you read quickly through the headlines of a newspaper – to see if there is anything worth reading – so scientists read the Contents page of a journal. As they read the title, they decide whether or not to read more. So it is worth giving a lot of thought to the choice of a good title to ensure that it attracts the attention of all those who might benefit from reading either the whole report or just selected parts (perhaps only the *Summary* or the *Introduction* and *Conclusions*).

Remember that the title of an internal report should be useful to all those who may see only the title – in a memorandum, or in a list of references in another document. Similarly, the title of a published article or report should be useful to those who see only the title, in another publication, as well as to those who have the whole report to study.

The title should be concise but unambiguous, and it should give a clear indication of the subject and scope of the work. It should include key words (words likely to be used in indices). For a published report you may also be asked to suggest additional key words that would facilitate information retrieval.

Bearing in mind its importance, the title should be reconsidered when your report is otherwise complete. Check that it is sufficiently direct and informative. Delete any superfluous words (for example: Aspects of . . .; A study of . . .; An enquiry into . . .).

For a printed report, include the following information as a footnote, but draw a circle around it to indicate to the printer that it is not to be printed: the number of folios (pages of typescript, tables, illustrations and other copy); a name and address for the editor to use in correspondence; and a short running title for the top of each printed page.

If your document is copyright, a statement to this effect (and the copyright symbol, ©) must be included in the document, normally immediately after the title page (on the title page verso: the verso page of an open book being the left-hand page, and the recto page the right-hand page).

The Abstract or Summary

Ensure that the *Summary* is complete, interesting and informative without reference to the rest of the report. Write in complete sentences, using words that will be understood by all those for whom your report is intended.

Although brief, the *Summary* must include a clear statement of the problem, and all your main findings, conclusions and recommendations – in the same order as in the report – because, apart from the title, this is all that some readers will actually read. The treatment of the subject may be indicated by such words as preliminary, detailed, theoretical and experimental. When experiments are reported the methods used should be mentioned. For new methods, the basic principles, range of operation, and degree of accuracy should be given. Statements of conclusions and inferences should be accompanied by an indication of their range of validity. If writing for a journal that publishes more than one kind of paper, the category to which your paper belongs should be mentioned (preliminary note, original paper, or review).

The *Summary* should be in the third person, in complete sentences, and in words that will be understood by all those for whom your report is intended, some of whom will have special interests very different from your own and may not understand all the technical details. It should enable the reader to decide whether or not to read more. For those who read only the *Title* and *Summary*, it should tell them as much as they need to know. For those who should read the whole report, it should capture their attention and provide a taste of what is to come.

In a published report, the *Summary* may be called an *Abstract*, because it may be extracted and used, with bibliographic details of the publication, by abstracting and indexing services. An *Abstract*, therefore, differs from the *Summary* of an internal report in that many readers will not have the report in their hands. Editors of journals state the maximum number of words to be used in preparing the *Abstract* (usually less than 300), and if you use more someone else may shorten your *Abstract*, and in doing so cut out things you consider important. Remember, also, that someone referring to your paper in a book or review article may reduce your conclusions to one sentence. Can you provide a suitable sentence in your *Summary*?

This section of the report, whether it is called an *Abstract* or a *Summary*, can be written only when your report is otherwise complete. Only then can you check that it contains: (a) everything you particularly want readers to know; (b) no information, ideas or claims that are not included in the report; and (c) no cross-references to tables, figures, or other pages of the report, or references to other publications.

The Table of Contents

If you think a Table of Contents would help your readers, list all the main headings, and perhaps also the sub-headings, with exactly the same wording and in the same order as in the report, and the page numbers. If you have used decimal numbers for headings, or for headings and paragraphs, these should also be included in the *Table of Contents*. Alternatively, if all paragraphs are numbered consecutively (not the headings), paragraph numbers should be used on the *Contents* page and in cross-references in the text, instead of page numbers.

The Introduction

Begin by stating the purpose and scope of the work, or your terms of reference. Include a clear statement of the problem (if there is one), any background information leading to the recognition of the problem, and – in an experiment – a clear statement of the hypothesis to be tested. Include a brief reference to any preliminary note and to other relevant investigations, by yourself or by others, to show how the work reported follows on from earlier work. Mention any new approach, any limitations, and any assumptions on which your work is based. If you have included an *Abstract* or *Summary* before this *Introduction*, do not repeat here things that should properly be in the Summary.

A clear, concise and interesting beginning may encourage readers to continue reading. As in the *Summary*, write in straightforward non-technical language, bearing in mind that some readers will read only the *Title*, *Summary* and *Introduction*, and any *Conclusions* or *Recommendations*. All readers should be able to understand those parts of the report in which they are interested, even if some parts can be understood only by specialists.

The Materials and Methods (or Procedure)

In a scientific or technical report enough detail should be included: (a) for readers to understand how your data were obtained; (b) to ensure that if the investigation were to be repeated by someone else, with appropriate experience, similar data could be obtained; and (c) to remind yourself, perhaps years later, how you did the work.

1   List the equipment used and draw anything that requires description (unless this is very simple).
2   State the conditions of the experiment and the procedure, with any precautions necessary to ensure accuracy and safety. However, when

several experiments are reported some details may fit better in the appropriate part of the *Results* section.

3   Describe your techniques or refer to an earlier report in which these were described. In particular, write the stages in any new procedure in the order of performance, and describe in detail any new technique or any modification of an established technique.

4   If necessary, refer to preliminary experiments and to any consequent changes in technique. Describe your controls adequately.

5   Include information on the purity and structure of the materials used, and on the source of the material and the method of preparation.

6   Use systematic chemical names or the pharmacological names recognised in the country in which your report is to appear. Give the internationally accepted name of the thing studied and refer to any factors that influenced your choice of material.

The Results

In this section provide a factual statement of your findings, supported by any statistics, tables or graphs derived from your analysis of data recorded in your investigation (or other diagrams that help you to present your results), but do not present information more than once (for example, in a table and in a graph).

The numerical data recorded on data sheets in a laboratory notebook, or on spreadsheets when using a computer, are not normally reproduced in a similar table when reporting the investigation. Representative data may be included, or all the data may be presented in a graph (as a scatter diagram), but any tables in the *Results* section should be summaries (results of your analysis of data). If your original data are needed by some readers (as in a thesis), they may be included in an appendix or made available in some other way.

You may describe representative successful experiments in detail; and it may be helpful to mention the unsuccessful experiments and wrong turnings that are part of every investigation.

Present the results as clearly and simply as possible, in an effective order (not usually the order in which the work was done), with enough words to give continuity and to help readers understand – but otherwise without comment. Take care not to start discussing your results in this section.

The Discussion

Consider the results presented in the previous section, with appropriate reference to any problem raised in the *Introduction*, to any hypothesis tested, and to relevant work by others. Refer to the theoretical background to your own practical work, to any limitations in your materials and methods, and to possible sources of error in the measurements made. List your main conclusions. Claim no more than can be substantiated from the results presented. Write in the past tense when commenting on what you did. Otherwise, write in the present tense.

You may be tempted to include much information sifted from the work of others, but each publication can usually be covered adequately in a few words (followed by a reference to the source). Some publications that seemed relevant when you made your notes may not be mentioned in your final report. The references cited should provide essential background for which you have no space in your report or should be needed for the development of your argument.

Most scientists have been misquoted or misunderstood. They may be pleasantly surprised when they are not. Before citing someone else's work, therefore, always make sure you have read the original publication and know exactly what was done, how it was done, and with what result. Do not rely on abstracts and reviews in which the original work of others may not be adequately or correctly represented.

When summarising other people's work try to preserve their meaning. This is not easy, and care is needed to ensure that readers do not take a different meaning when words are repeated out of context. If you need to quote someone else's exact words, ensure that all the words and punctuation marks are copied correctly, and make clear that you are quoting verbatim, either by using quotation marks (as on page 13) or by indentation and an acknowledgement (as on page 48).

The Conclusions

Your *conclusions* may be listed at the end of the *Discussion* or after a separate heading. They should follow from arguments and evidence included in your report, and provide an effective ending. They should be numbered, to ensure that they are in order and distinct; and each conclusion should be a precise and concise but clear statement.

The Recommendations

If it is within your terms of reference to make recommendations, these should be practicable and should arise directly from your conclusions. They, too, should be listed as separate, numbered statements advising, for example, precisely what should be done, when it should be done, and by whom.

The Acknowledgements

If anyone helped you, with either the work reported (with materials, assistance or advice) or the preparation of the report, this should be acknowledged simply and concisely. It is normally sufficient to write 'I thank . . . for . . ., and . . . for . . .', making clear who contributed and what they did. It is advisable to let anyone mentioned in this section read what you have said about them, so that they have the opportunity to comment.

It is not normally necessary to thank colleagues whose contribution was a routine part of their employment, and was insufficient to merit their inclusion as co-authors, but you may be required to state the source of finance – for the work and for the report.

Some authors add that they take responsibility for the final arrangement, the opinions expressed, and for any remaining mistakes. In scientific writing this truism should be omitted, but some statement may be required, for example by a firm or a government department, to indicate that the views expressed are not necessarily officially endorsed. This is another occasion when the house rules must be followed.

Copyright.   In a report on an original scientific investigation the question of copyright does not usually arise; but before reproducing copyright material you must obtain such permission as is required by law. For further advice on copyright, see the *Writers' and Artists' Yearbook* (A. & C. Black, London), but remember that there are differences in law in different countries and consult your editor or publisher if you are writing a review or a book. If quotations are included for purposes of criticism or review, or if tables or illustrations are modified, the permission of the copyright holder may not be necessary. A proper acknowledgement of the source of quoted material (as on page 13) or of the data upon which an illustration or table is based (as in Table 9.3 and Figure 9.4) may be all that is required.

However, anyone wishing to reproduce copyright material should write both to the owner of the copyright and to the author (or publisher) of the work in which the material first appeared. In seeking permission to reproduce material, lines to be quoted should be identified by the title of the work, the date of publication (and the number of the edition and volume), the page number, and the number of the lines on which the quotation starts and ends, with the first few and the last few words of the quotation. Illustrations and tables which are to be copied should be identified in a similar way, but by their number and by the number of the page on which they appear.

Prepare three copies of this letter with a statement below your signature in the form of a reply. This should state that permission to use the above material in the way described is granted. There should then be spaces for a signature and the date. Send two copies, with a stamped, addressed envelope, so that the copyright holder can return one signed copy and retain the other as a record. The copyright holder may require a fee, and may state how your acknowledgement of the source of this material is to be worded.

The Bibliography or List of References

Relevant previous work may be mentioned in the *Introduction, Methods,* and *Discussion* sections, but not in other sections, with complete bibliographic details of each publication listed in a *Bibliography* or list of *References,* as appropriate.

Use the heading *Bibliography* if your list includes bibliographic details of published works that you consulted in preparing your report, or that have influenced your thinking, but are not necessarily cited in your report. A bibliography may also include annotations. A note after the heading should state the principles on which a bibliography has been compiled.

Use the heading *References* if your list of sources of information or ideas comprises complete bibliographic details of every publication cited in your report, but no others, as is usual in most scientific and technical publications.

The way in which bibliographic details are listed must be consistent with your house rules, if there are any. Otherwise, look at a recent internal report to find out what is acceptable to your employer, or at a recent issue of the journal in which you hope to publish your work. See also, *Citing sources of information* (page 129).

References may be listed in alphabetical order (see British Standard BS 1749) according to the surname of the author (or editor) or that of the first of a number of authors (or editors) or in numerical order, depending on how you have cited sources in the text. An editor (ed.) or translator (trans.) is indicated by an abbreviation after the name. Recommendations for bibliographical references are also the subject of British and International Standards (BS 1629 and ISO 690).

Care is needed in checking the accuracy of all references, including the spelling of proper names, because each reference is both an acknowledgement of someone else's work and a source of information for the reader.

The Appendices

Details that would be out of place in the body of a report, but which may be required by some readers, for example tables of original data, may be included in an *Appendix* or made available in some other way.

The Index

If an *Index* is needed, it can be prepared only when the typescript is compete and the page numbers are known (see *Preparing the index*, page 155); and for a printed report it must be prepared from the page proofs.

The Distribution list

All those who are to receive copies should be listed, in alphabetical order, either on the Title Page or at the end of the report, to ensure that copies are sent only to those who require them. If necessary, a memorandum can be sent to others who may be interested to inform them of the report's existence.

Theses and students' project reports

Theses

The word thesis means a statement, proposition or position which a person advances and is prepared to maintain. The word is also used as a synonym for a dissertation: a written presentation of a subject, a contribution to knowledge, usually prepared by a candidate for a higher degree.

The purpose of a thesis is to train the mind of the writer and to show how far it has been trained. The writer, after years of thought and study, must have mastered the subject, and must convey the ideas and understanding that come from observation, reading, discussion and reflection.

The thesis for a Master's degree is based upon a training in the problems and methods of scientific investigation: upon *independent* research. The thesis for a Doctorate is based upon independent *original* research: upon an investigation in which the frontiers of knowledge have been explored and extended.

The limited purpose of a thesis is to contribute to the solution of a problem. Great care is needed, therefore, in selecting a problem to which a contribution may be made in the time available.

The parts of a thesis are the same as for other scientific reports (see pages 133–4). However, because it has to satisfy an examination requirement, the thesis has a restricted readership: it is written for the few specialists who will judge its merit.

The Title Page should include the full title (and any sub-title); the full name of the candidate; the qualification for which the thesis is submitted; the name of the institution to which the thesis is submitted; the department, faculty or organisation in which the research was conducted; and the month and year of submission.

The Title Page should be followed by a Table of Contents and a List of Tables and Illustrations. The *Acknowledgements* should include a declaration in which a note is made of any material in the thesis which has been used before, and of the author's part in any joint work included in the thesis. The *Summary*, of about 300 words, should state the nature and scope of the research and of the contribution made to knowledge of the subject. A brief summary of the method of investigation may be appropriate, an outline of the major divisions or principal arguments of the work, and a summary of any conclusions.

After a short introduction, a concise critical literature survey of relevant previous work may be required (as a separate section). This gives perspective to the work and shows existing knowledge as the basis of further discoveries. However, candidates should refer only to those things that they can discuss intelligently in their oral examination.

The *Materials and Methods* and *Results* sections should show the approach to the problem, and what has been added; and in the Discussion section the results should be interpreted and discussed in relation to previous work. The references cited must make clear the writer's understanding of relevant literature in the general field, and of all references directly concerned with the problem. In all its parts the thesis is a measure of scholarship and industry as well as of research ability.

The thesis must meet the detailed requirements of the examining body to which it is to be submitted (see also BS 4821 on the *Presentation of theses*) and it should normally be similar in form and arrangement to theses previously submitted for the same qualification in the same Department. However, many theses are too long and contain more tables and illustrations than are necessary. Subject to the supervisor's agreement, the body of the thesis (the part following the critical survey of previous work) may be prepared in a form which is suitable for immediate submission to an appropriate journal, as one or more research papers. If the original numerical data are included they should be in an Appendix.

Students' project reports

Project work, which is part of many degree courses in science and engineering, involves the preparation of a report that is more demanding than anything a student has prepared previously. If the project is based on a literature search and reading, with no opportunity for personal observations, it may be called a dissertation, an extended essay, or a review. If the project includes an investigation, with the student's own observations related to relevant work by others, the project report should be written as a scientific paper with the usual headings (as on page 133). Any special requirements should be included in detailed *notes for guidance* issued to the students taking a particular course.

A project report will have a concise title, and the work as a whole should be based on clear *terms of reference*, stated in writing so that there is no possibility of misunderstandings and provided by the supervisor (or agreed by student and supervisor) before the investigation begins.

In choosing a problem to investigate, and in planning the work, the student should be supervised, to ensure: (a) that the project is one that can be completed in the time available; and (b) that the literature search and any practical work begins well.

Preparing a project report is a test of a student's ability: (a) to demonstrate knowledge and understanding and critical evaluation of relevant published work; (b) to communicate information and ideas in writing, supported, if appropriate, by tables and diagrams; and (c) to complete an investigation – including the report – on time.

Project assessment.   It is difficult to arrive at an objective assessment of extended practical work, but easier to make a subjective judgement based on the student's general attitude to the work and to assess the quality of the project report. For an external examiner, unless there is an oral examination, the project report is the only guide to the quality of the work.

A good student will be expected to write a clear, concise, considered, critical, balanced, and well-organised written report. In contrast, an incomplete, opinionated, superficial or uncritical composition, with inadequate reference to relevant sources, will be recognised as lacking in rigour.

However, even from the report it may be difficult for an external examiner to decide how much of the report is the student's work and how much is the supervisor's.

Different projects involve different methods of investigation; some require more background reading than others; and some provide more scope than others for the student to show originality, initiative and ingenuity. These differences, which make objective assessment difficult, must be carefully considered if all students are to be treated fairly.

There is probably no one solution to these problems, but it is possible to list things that should influence the final assessment. These cannot all be judged at the end of the work.

1   The student's ability to define the problem to be investigated, or to state the purpose of the work, if the subject was chosen by the student.
2   The student's ability to plan the procedure, make precise measurements and prepare accurate records on well-organised data sheets.

3   The student's ability to evaluate published work, or personal observations, to analyse numerical data (as appropriate), to argue logically, and to draw valid conclusions.

4   The student's ability to relate new observations and results to knowledge of the subject derived from the published work of others.

5   The thoroughness with which the work was tackled in relation to the time available.

6   The student's ability to select relevant material and reject what is irrelevant.

These aspects of the work might be considered of comparable importance and given equal weight in a marking scheme. But, whatever the method of assessment, students should know, before work on the project is started, how the project report is to be presented and how the project as a whole will be assessed.

13 Writing a report on your investigation

Write from the start

Preparing your manuscript (first draft)

If you hand-write your first draft, start each section with a heading or sub-heading on a new sheet of paper, write each paragraph on a separate sheet, and prepare each table, graph or other diagram on a separate sheet. Then, as when using a word processor, it is easy to rearrange your material or make additions or deletions.

You are also advised to write on alternate lines of wide-lined paper, or on unlined paper, so that you have plenty of space for additions and corrections, and to use carbon paper so that you have a copy of each sheet that you can keep in a safe place, separate from the copy on which you are working. Alternatively, when word processing, save your work regularly and ensure that you always have an up-to-date copy on a back-up disk in a safe place.

First prepare a Table of *Contents* based on your topic outline. Then prepare drafts of the Front Cover, a distribution list and an *Introduction*. This will help to focus your attention on your readers, why they require this report, and how it is to be organised so as to satisfy their needs.

Write the *Methods* section next, as soon as you have standardised the procedure to be used in collecting any original data, while the details and any difficulties are fresh in your mind. Then accumulate material for the *Results* section, and prepare any tables or diagrams, as you collect and analyse your data. By preparing these tables and diagrams before actually writing the *Results* section, in which they are to be mentioned, you can avoid repeating in the text information that is already presented in such illustrative material. If necessary, you can also repeat any observations, and obtain more data while the materials and equipment you require are still available.

Although you cannot write the *Discussion* section or any conclusions until the *Results* section is complete, it is essential to note relevant points under appropriate sub-headings, throughout the work, as they come to mind – so that they are not forgotten.

In other words, you are advised to work on the report as a whole throughout any investigation (see Figure 12.1), so that as you write and revise your draft it is always an up-to-date progress report. This should be easier than collecting all relevant information first, and then having to write the report at the end, near your deadline, as a distinct and separate task. However, the *Summary* or *Abstract* cannot be written until the work is otherwise complete. Then the *Introduction* should be reconsidered, and your draft of the whole report reconsidered and revised.

Improve your writing

When your manuscript is complete, whether it is handwritten or word processed, think of it as a first draft. Read it and correct any obvious mistakes. Then, if you have time, put it on one side while you get on with other work. One way to do this, is to ask a colleague to read it and to let you have any comments or suggestions for its improvement. See *Revising*, pages 46–7).

When you read your report again, after a break of a few days, or longer if this is practicable, you will see things in a fresh light. For example, you will find statements that are ambiguous or could be better expressed, and sentences and even paragraphs that are out of place. If it is handwritten, with a separate sheet for each paragraph, you will find it easy to add, delete, or change the order of paragraphs, if you need to.

Checking your manuscript (first draft)

It is not possible to check your manuscript thoroughly by reading it through once or twice. Instead, check one thing at a time:

1 Is the Title Page complete (see page 139)?
2 Does the title provide the best concise description of the contents of your report?
3 Is the use of headings and sub-headings consistent throughout the report; are the headings concise; and are all the headings and sub-headings used in planning the report still needed?
4 Is the Contents page still needed? If it is, are the headings identical with those used in the report?
5 Is the purpose and scope of the report stated clearly and concisely in the Introduction?
6 Have you achieved your purpose and kept within the terms of reference?

7 Has anything essential been left out? Have you answered all the reader's questions (see page 41, and 133)? Are your conclusions clearly expressed?

8 Is each paragraph relevant, necessary and in its proper place? Are the paragraphs in each section in the most effective order? Is the connection between paragraphs clear?

9 Is each paragraph interesting? Is the topic clearly indicated and is everything in the paragraph relevant to the topic? Is the emphasis in the most effective place?

10 Are all arguments forcefully developed and taken directly to their logical conclusion, and is anything original emphasised sufficiently?

11 Is there an important point that could be more clearly expressed, or made more forcefully in an illustration? Should any illustration be replaced by a few lines of text?

12 Is each statement accurate, based on sufficient evidence, free from contradictions, and free from errors of omission? Are there any words such as *many* or a *few* that can be replaced by numbers?

13 Are there any faults in logic or mistakes in spelling or grammar?

14 Is each sentence necessary? Does it repeat unintentionally something that has been better expressed elsewhere?

15 Could the meaning of any sentence be better expressed? Are there any unnecessary words?

16 Is each sentence easy to read? Does it sound well when read aloud, and is the emphasis in the most effective place?

17 Are any technical terms, symbols or abbreviations sufficiently explained?

18 Are all the words to be printed in italics underlined (see page 154), and are those to be in bold underlined with a wavy line? Or, if appropriate, are they already in italic and bold print?

19 Are you consistent in spelling, and in the use of capitals, hyphens and quotation marks?

20 If you added anything as a footnote, while you were preparing your manuscript, check that this material has been incorporated in the text. A footnote may be required on the Title Page (see page 139); and footnotes are essential in some tables (for example, see Table 3.1). Do not use footnotes for information that ought to be in the Acknowledgements or in the list of References.

21 Are all the references accurate, especially the spelling of proper names? Do the dates in the list of References (on your index cards) agree with those given in the text?

22 English is a language of international communication. If your report is for a wide readership, or for readers with different interests, check that your prose is clear and direct.

23 Is each table and each illustration referred to, by its number, in the text?
24 Are all your revisions improvements? Is every word, letter, number and symbol in your manuscript legible?
25 Are all the pages numbered and in their correct order?
26 Are the headings in the text identical with those in the Table of Contents.
27 If the paragraphs are numbered, have you included these numbers in cross-references in the text and on the Contents page? For a printed report, if all the paragraphs are numbered, cross-references can be included in the typescript – but if, as is more usual, only the pages are numbered, cross-references must be added at the proof stage unless you are preparing camera-ready copy.
28 Does the revised report read well and is it well balanced?
29 Check your *Summary* (see page 140).
30 Check the Acknowledgements. In particular, have you obtained written permission to use any copyright material?

Preparing your typescript

If your report is word processed by someone who has not previously prepared similar documents, emphasise that normal office rules for correspondence do not apply. Give clear instructions (which must be consistent, for example, with your organisation's house rules for internal reports, the regulations of the awarding body for theses, or the notes for guidance issued by the editor of the journal to which your paper is to be submitted), such as the following:

1 Date typescript required.
2 Use A4 paper (210 × 297 mm).
3 Number of copies required.
4 Use Times New Roman (a *serif* font): 10 point for single spacing, or 12 point for one and a half or double spacing. A *sans serif* font (for example, Arial) may be preferred for headings.
5 Print on one side of the page only.
6 Leave a 40 mm margin on the left; and about 25 mm on the right, top and bottom of the page.
7 Do not justify the right-hand margin; and do not use hard returns. Do not insert a hyphen at the end or at the start of a line. Use hyphens only in words that must be hyphenated.
8 Use two hard returns at the end of a paragraph, and do not indent the first line of the next paragraph.

9 Number the pages at the bottom centre (to leave space for the alphanumeric reference, if required, at the top right-hand corner of each page) or, for a report that is to be printed, number each page at the top right-hand corner.

10 If the report is to be printed, include the surname of the first author at the top left-hand corner of each page.

11 Use a separate page for each table, with at least 40 mm margins. Type the number of the table and the heading immediately above the table. Do not end the heading with a full stop unless it is a sentence. For a printed report include the tables at the end of the typescript, before the legends to figures.

12 For a printed report, leave spaces in the typescript for any mathematical expressions or chemical formulae that are to be type-set by the printer.

13 Centre section headings (marked A in the margin of the manuscript) at the top of a new sheet; shoulder sub-headings (marked B) with a line to themselves; and shoulder minor headings (marked C) and, after a full stop, continue on the same line with the next sentence.

14 Use upper case only for the initial letter of each sentence, heading or proper noun.

15 Underline only those words underlined in the manuscript (or, if required by house rules, type them in italics): the titles of publications (see page 195); the scientific names of organisms (generic names, for example, *Homo*; and species names, for example, *Homo sapiens*); words from a foreign language that are not accepted as English words (for example *modus operandi*) and abbreviations of such words (see *Abbreviations*, page 64); and the words *either* and *or* when it is necessary to emphasise an important distinction. All these are words that in a publication would be printed in italics. To a printer underlining means 'Print in italics', and underlining with a wavy line means 'Print in bold type'. When word processing a document for internal circulation or for publication, the use of underlining, italics and bold may be determined by the employer's or publisher's house rules. Note that a heading attracts enough attention if given a line to itself, so it is not usual to underline a heading unless it ends with a full stop and the text is run on (on the same line).

16 Type the Contents pages when page or paragraph numbers are known. If the paragraphs are numbered, cross-references in the text can be added in the manuscript – but if, as is more usual, only the pages are to be numbered, both the page numbers and the cross-references must be added either after the typescript has been checked or they must be added to the proofs.

17 For a printed report, list the legends to the figures at the end of the typescript, after the tables, below the heading Legends to Figures. Note that a concise legend, if it is not a sentence, should not be followed by a full stop.

Checking your typescript

1 Compare the typescript with the manuscript, to ensure it is a complete and accurate copy. Mark any corrections or amendments on one copy of the typescript. File your manuscript: do not discard it.
2 Does your report read well? Is it well balanced?
3 Are there any typing errors, or mistakes in spelling or grammar?
4 Are all dates, numbers, and mathematical and chemical formulae correct?
5 Are all the references to tables and figures in the text numbered correctly?
6 Is the spelling of all technical terms and proper names correct?
7 Check the wording and punctuation of all quotations and references against the original. If words are omitted from a quotation the gap should be indicated by three stops . . . and anything added should be in [square brackets].
8 Are all references cited in the text up-to-date and in the list of References? Read the papers cited again to make sure you have taken the right meaning.
9 Are the headings to all tables and the legends to all figures adequate?
10 Is the source of any quotation, table or figure properly acknowledged, and where necessary has the written permission of the copyright owner been obtained?
11 Have any diacritical marks (in quotations from other languages) and symbols been inserted correctly?

Preparing the index

Read one copy of the typescript (or page proofs), marking in a conspicuous colour all words to be included in the index (topic words). Then go through the report, page by page, writing each word so marked on a separate index card with the number of each page on which the word is underlined. Keep the index cards in alphabetical order (see BS 1749 *Alphabetical arrangement*) to facilitate the addition of page numbers. The index can be typed from these cards. Alternatively, if a word processor is used, you can indicate in the text words to be included in an alphabetical index. Either way, you have to decide which words to include and then, in each entry, direct the readers' attention only to those pages on which the word is defined, explained or discussed – not to every page on which the word is used.

Sub-entries should be indented and arranged in alphabetical order below the relevant main entry. For a printed report, unless the publisher specifies otherwise, each main entry and sub-entry should start on a new line; the first page number should be preceded by a comma and successive page numbers should be separated by commas. When an entry refers to the main subject considered on successive pages, only the first and last page numbers should be given, joined by a dash. No punctuation is used at the end of a main entry or sub-entry. If the publisher specifies that sub-entries are to be run on, separate them by semicolons.

Cross-references may be useful. Alternatively, the same page number should be included under different headings. *See also* entries, at the end of an entry, may also help the readers.

In a typed report the index should be in single spacing, but in typing an index for a printer use double or treble spacing and leave wide margins. The pages that include illustrations (or definitions) should be underlined (if you would like them printed in italics) or underlined with a wavy line (if you would like them printed in bold). A note should be included at the start of the index to explain that the pages with illustrations (or definitions) are printed in italics (or bold) as appropriate. To avoid confusion with page numbers any dates in the index should be in parenthesis (in round brackets). Keep a copy of the index with your copy of the typescript.

Recommendations for the preparation of indices are the subject of BS ISO 999 and of ANSI/NISO TR-02.

Marking the typescript for the printer

If your report is to be printed, occasional words may be corrected in ink between the lines of the typescript (but not in the margin). However, most publishers prefer to receive a word-processed typescript that contains no handwritten corrections or amendments. They will probably also ask for a copy on disk (with a note of the make of and model of the computer used, details of the word-processing software and operating system used, and a list of file names and their contents).

Check that each folio (sheet of typescript or other copy) is numbered correctly (top right-hand corner) and that the surname of the first author is also given (top left-hand corner). If any folios are added later, the two preceding folios should be marked, for example: 29 (folios 30 a-c follow) and 30a (folio 30b follows) and 30b (folio 30c follows) and 30c (folio 31 follows). If any folio is removed, the preceding folio should be renumbered (for example, if folio 12 is removed, folio 11 is renumbered folio 11–12).

It is essential to check the typescript and the illustrations carefully. Only printer's errors should be corrected in the proofs: authors should not ask for changes at this late stage. Indeed, for some publications authors may be asked to provide camera-ready copy so that a typescript can be published without the need to provide the author with proofs for correction of printer's errors. If camera-ready copy is required, the editor of the publication will provide detailed instructions.

Where necessary, include marginal instructions for the printer on the typescript. For example, explain any unusual symbols or Greek letters. Underline only words or symbols to be printed in italics, if these are not already in italics. If anything that is correctly typed could be considered to be a mistake, write 'Set as typed' next to the words or letters. Use marginal letters to indicate grades of heading (see page 136). Indicate the position of each table and illustration by a marginal note in the text.

Photographs (see also page 107) should be black and white and should normally be full plate or half plate. When several photographs are to be included in the same plate, prepare a key for the printer to show the arrangement required. Do not mount the photographs, unless asked to do so by the editor.

If only a part of a photograph is required, this part should be marked by a rectangle on a transparent overlay. Alternatively, prepare an enlargement from the relevant part of the negative. Any other information required by the printer should be marked lightly on the reverse of the photograph, preferably in the margin. Care is needed in writing on the back of a photograph (or on an overlay) as lines may show on the photograph and spoil the plate – as may an over-inked rubber stamp on the back or pressure marks caused by paper clips.

When lettering, a scale, or other marks have to be inserted by the printer, copies of the photographs with the necessary additions should be provided or the additions should be printed on a transparent overlay, according to the requirements of the printer. If any illustration is without letters or numbers, or some other clear indication of its correct orientation, the word *top* should be written lightly on the reverse (preferably in the margin).

Corresponding with an editor

If your report is to be published as a scientific paper, do not publish prematurely. Submit a paper for consideration by an editor only if it is original and includes new findings (and/or new interpretations supported by sufficient evidence). Before submitting your paper, consider which journal would be most appropriate. Do not submit it to more than one journal at a time, and do not submit a typescript if it has already been published or accepted for publication elsewhere.

Send your typescript to the editor at the address given in a recent issue of the journal. The typescript (including the title page, text, references, tables, and legends to figures) and the artwork should be kept flat with stiff cardboard and posted in one envelope. The pages should be held together by a paper clip (not by a staple) or they should be punched and threaded on a treasury tag. For a long report, one paper clip may be used for each part and an elastic band put around the whole typescript. If the editor requires more than one copy of the typescript (and an identical copy on disk, see page 190), all copies should be sent in the same envelope.

The editor will acknowledge receipt of your typescript. Then there will be a delay while it is sent to one or more referees – who will comment as to its suitability for publication in this journal. You can save yourself time and help the editor and referees if you consider the following questions yourself, before submitting your work to an editor.

A check list for referees (and authors)

1 Is the paper suitable for publication in this journal?
2 Do you recommend publication of the paper: (a) as it is; or (b) after revision?
3 Is the work reported original; and has any part been published?
4 Is the work complete? Is it a contribution to the subject?
5 Are there any errors, or faults of logic?
6 Are there any ambiguities? Are any parts badly expressed? Are any parts superfluous? Are any points over-emphasised or under-emphasised? Is more explanation needed?
7 Does the typescript conform to the journal's requirements, as indicated in the notes for authors?
8 Should all parts of the paper be published?
9 Is the title clear, concise and effective?
10 If key words are required, are those suggested appropriate?
11 Is the abstract comprehensive and concise?
12 Are the methods sound? Are they described clearly and concisely?
13 Are the illustrations and tables properly prepared?
14 Are any conclusions supported by sufficient evidence?
15 Are all relevant references cited? Are any of those cited unnecessary?

Even if the editor wishes to accept your paper, improvements are likely to be suggested. The editor speaks from experience and any comments are based on the confidential reports of referees. If you do not like them, do not reply immediately. Write a reasoned reply when you are ready to submit a revised paper. Referees may be wrong, but you should welcome their comments. If they have misunderstood, others might misunderstand. If they were not convinced, others might not be convinced. So take the opportunity to think again, to correct any mistakes, to clarify any difficult or ambiguous points, and to consider other revisions. You will probably find that you are pleased to have had the opportunity to look afresh at your typescript.

In returning your revised typescript to the editor, say how it has been improved. If any of the referees' suggestions have not been accepted, say why not. Responsibility for the typescript rests with you, just as responsibility for its acceptance or rejection for publication in any journal rests with its editor.

Some journals receive for consideration many more papers than they could publish. Rejection, therefore, does not necessarily mean there is any-thing wrong with your paper. Perhaps the editor will suggest another journal that may be more appropriate. Sometimes one editor rejects a paper, the importance of which is recognised by another editor. However, if a paper is rejected take the opportunity to think again, to see if it can be improved, before you revise it to conform to the house rules of another journal.

Checking the proofs

If your work is to be published, proofs will be prepared from your typescript. These will be sent to you for checking, and so that you can prepare an index (if one is needed). Any printer's errors should be corrected in red ink. Alterations should not be made at this stage. However, if you must make changes, any additions or deletions (in black or dark-blue ink) should be matched by corresponding deletions or additions, of words or phrases of the same length (counting each letter and each space).

1 All notes for the printer and any corrections must be marked on the proofs, not on the typescript.
2 Corrections must be indicated clearly for the printer, in the right and left margins and with appropriate marks in the text (see *Copy preparation and proof correction* in the list of Standards in Table 11.1). Words deleted should be crossed out by a horizontal line, and letters by a nearly vertical line. Any marginal comments or instructions for the printer, which are not to be set in type, should be preceded by the word PRINTER.
3 The questions asked by the printer, usually marked by a question mark in the margin, must be answered carefully.
4 Write the numbers on the Contents page and in cross-references in the text; and prepare the index.

5 Ask someone to read the typescript aloud while you check that the proofs are an accurate copy.
6 Read the proofs several times to check for printer's errors and for mistakes in spelling.
7 Check the accuracy of all dates, numbers and formulae.
8 Check the spelling of all specialist terms and proper names.
9 Check the wording and punctuation of all quotations and references against the original.
10 Check that the tables and figures are in the right place, that they have the right headings and legends, and that the numbers used in cross-references in the text are correct.
11 Check the illustrations to ensure that each is a good copy of the original, that all lines are good, and that there are no extraneous marks.
12 Keep one copy of the corrected proofs and return one copy to the editor.

Summary

How to prepare a report on an investigation or a paper for publication

This check list is a summary of the procedure recommended in this chapter together with relevant advice from other chapters. For an internal report, follow the stages marked by numbers.

> Follow the additional stages, inset, when preparing a paper for publication.

1 Keep full bibliographic details of every relevant reference consulted. See *Making notes as you read* (page 127), and *The list of References* (page 145).
2 Keep a copy of everything you write, in a safe place, separate from the top copy.
3 Write the title for the first draft and your Introduction, and prepare a provisional list of Contents with headings and sub-headings before you start work. Revise these as necessary during the investigation.
4 Keep a day-to-day record of your work in a laboratory notebook (see pages 10–11).
5 Write the Materials and Methods section as soon as your procedure has been established and any initial difficulties have been overcome.
6 Prepare tables with effective headings, and drawings and diagrams with concise but complete legends, as each observation or experiment is completed.

See *Preparing tables*, page 95.
Preparing graphs and charts, page 97.
Illustrations contribute to clarity, page 105.
Writing the legend, page 115.

7 Prepare notes on each observation or experiment as your work proceeds.
See *Writing helps you to observe* (page 9)
Writing helps you to think (page 13).

8 List points arising from your work and from relevant work by other people that you must remember when writing your Discussion.

9 When your work is complete prepare a detailed topic outline.
Should your work be published?
If there are new findings, should your work be published in whole or in part? Should it be published as one or as more than one paper? Which journal(s) would be most appropriate?
Read and obey the *Instructions to authors* for the journal(s).

10 Write the first complete draft.
See *Writing*, page 44.
How to write your report on an investigation, page 132.
Preparing your manuscript, page 150.

11 Revise the first complete draft.
See *Revising*, page 46.
Checking your manuscript, page 151.

12 Ask two people to read the second draft (see page 47); and then revise your manuscript in the light of their comments and suggestions.
Check that publication is acceptable to your supervisor and employer and that nothing to be published is classified as confidential or secret. See also, patents (page 47).

13 Obtain written permission to use any copyright material (see page 144).

14 Read all the references cited in the text to make sure that the work of others is correctly represented and that the bibliographic details are correct on your index cards.

15 Make sure that everything in the manuscript is in the right place.
For a printed report place the tables after the list of references, not interleaved in the text.

16 See advice on *Preparing your typescript* (page 153).

17 Prepare the list of References from your index cards, including only works cited in the text.

18 Prepare the illustrations (see page 110).
Prepare the Legends to figures (see page 115), and put these at the end of the typescript, with your artwork, after the tables.

19 Check the typescript (see page 155).

Check that the typescript meets all the requirements listed in the journal's *Instructions to authors*.

Minor corrections, additions and deletions should be marked clearly on all copies of the typescript and/or on the disk if a disk is submitted as well as or instead of the typescript.

20 Obtain clearance for the corrected typescript from your supervisor, head of department, or employer, as appropriate.

Mark the typescript for the printer, and keep a copy.

Send the typescript (as hard copies, on disk, or by e-mail, as required in the *Instructions to authors*) to the editor of the journal, with a short covering letter. See *Corresponding with an editor* (page 157).

Correct the proofs (see page 159), then return one copy of the corrected proofs to the editor and keep one for your records.

14 Talking about science

Preparing your talk

The expression *reading a paper* is misleading, because speech is not the same as writing (see page 72). If you are to give a talk that will also be published, it is best to prepare the typescript for publication and then to speak from notes. Listeners cannot assimilate all the detail that goes properly into a written report but which has no place in a talk.

A *talk* or *lecture*, delivered before an audience, is an opportunity, for example: to provide a foundation for independent study or research (to introduce); to present a subject and view it as a whole (to stimulate interest); to present facts and opinions not readily available elsewhere (to inform); to develop an argument (to persuade); or to draw attention to important points, contradictions and uncertainties (to stimulate further thought).

A *presentation* is a special kind of talk, an exercise in persuasion involving one or more presenters, in which something new is presented to an audience for consideration. This could be an idea, a policy, a document (for example, a report introduced on the day of its publication), or a product (for example, new equipment promoted on the occasion of its becoming available to view, to order or to buy). Each presentation should be complete in itself, but should leave the audience interested, impressed, and wanting to buy or to know more.

A poster presentation at a conference, or as part of a business meeting, is an opportunity for someone who is not giving a talk to present, for example, information about a new idea, a new procedure, or a new item of equipment, or to summarise work in progress. Alternatively, a speaker may use a poster, after a talk, to provide more detail than would have been appropriate in the talk.

Giving a poster presentation

Give your poster an eye-catching title, at the top, in large letters. Use large letters for other words you want people to see at a distance, and indicate sequences by large numbers and/or arrows. Place any diagrams or photographs together if they are to be compared; and ensure that anything requiring close examination is at eye-level. Be brief: resist the temptation to include more information than can be understood by non-specialists in two or three minutes. Stand next to the poster so that you can answer questions or discuss your work with anyone who would like to know more. If appropriate, prepare a handout for those who require more detail, and leave copies in a pocket attached to the poster so that they are available when you are not present to answer questions.

Talking to an audience

Whether you are speaking alone or as part of a team giving a presentation, before agreeing to talk on any subject you need to know: (1) what exactly you are asked to speak about (a title or precise terms of reference); (2) who will be your audience; (3) why they would like you to speak to them; (4) when; (5) for how long; (6) where (the place, the size of the room and the facilities available).

If you are a member of a team making a presentation, you will also have to understand what each member of the team is to contribute, and have at least one trial run to ensure that each contribution can be completed in an agreed time, that the different contributions are in an effective order and are well co-ordinated, and that the whole presentation runs smoothly and ends on time.

In seeking employment you may have to give a short talk as part of an employer's selection procedure. In employment you may be asked to give instruction as part of a training course, or to organise a presentation relating to a new product or new requirement. Even in such different situations, the qualities required of a speaker are the same: enthusiasm, simplicity in the use of language, and sincerity.

To help you to relax, if you feel apprehensive about talking to an audience, speak up if you are not using a microphone but use the same voice as in conversation, the same gestures, and the same pauses, so that you move forward at the same pace – unhurried – maintaining eye contact with everyone present. Ronald Reagan (1990) in *An American Life* gave five rules about public speaking: (1) Use simple language. (2) Do not use a word with two syllables if a one-syllable word will do. (3) Prefer short sentences. (4) If you can, use an example: an example [or an analogy] is better than a sermon. (5) Audiences are made up of individuals, so speak as if you were talking to a few friends.

People can proceed at their own pace when reading. If something is not clear immediately, they can stop and try to work things out. But if listeners are trying to understand what has just been said, they will not be concentrating on what is being said next. That is to say, in a talk everyone must understand all that is said – at first hearing. The speaker must ensure, by adequate preparation, that all is right from the start.

Analysing your audience.   As in writing, consider your audience. What are their interests? What are their likely feelings about the subject of your talk? What do they need to know? How well do they know you? What do they expect of you? How do you expect them to benefit from your talk? That is to say, what is your purpose in giving the talk? Do you intend, for example, to encourage, entertain, explain, inform, inspire, instruct, or persuade? What can you achieve in the time available? Consider the following advice.

Designing your message.   There are many ways to begin (see pages 42–3 and *How to begin*, page 81); but never begin by saying that you are not really qualified to speak on this subject. Decide about this before you agree to talk. Having agreed to speak, do any necessary background reading. Make sure that you do know enough about the subject. You must be self-confident if you are to retain the confidence of your audience.

If you have prepared a written report on the subject of your talk, remember that speaking is not the same as writing. A good composition, prepared for silent reading, will not make a good talk if it is simply read aloud. If you must read your talk, write it so that it will sound well when read aloud.

In writing you might explain: 'This is what was done . . .'. But in a talk you would use the first or second person: 'I did this . . .', 'We did this . . .' or 'As you know . . .'.

In scientific writing most people avoid the colloquial language that is natural in conversation. If in a talk you follow the advice given here, and speak as you would to a small group of friends, you will use colloquial language.

In scientific writing there is no place for rhetorical questions. But have they a place in a talk? Can you use a rhetorical question to make listeners think about what you have just said, or to start thinking about what you plan to say next? Are there other things discouraged in writing that should be encouraged in speaking?

Communicating your purpose.   Whereas the reader does not require your plan (see page 43), listeners do need a map or guide to help them find the way. Some introductory phrases that would be superfluous in writing may help listeners to understand how your talk is organised, and how it is progressing. For example: 'As the title of my talk indicates, . . .'. 'So far we have seen that . . .'. 'The next thing I want you to consider is . . .'. 'As I have already emphasised . . .'. And, when you are sure you are about to end, you could say: 'To summarise, . . .', or 'In conclusion, . . .', or 'I leave you with this message . . .'.

Also, apart from the way you choose to express your thoughts, remember that listeners cannot assimilate all the detail that is needed in some written reports, but has no place in a talk. Most inexperienced speakers, and many experienced speakers who have not taken enough trouble in planning their talks, attempt to cover too many main points, use too many visual aids, and include too much supporting detail. As a result, little of what is said is likely to be remembered, and few of those present will be able to recall accurately even the main points – unless they made notes.

Repetition is usually undesirable in writing, because the same words can be read again, but repetition helps listeners remember what is said in a talk. You may, for example: (a) state the title of your talk; (b) say briefly what each part of your talk is to be about; (c) state the main point you want to make at the start of each part of your talk; (d) explain each main point with some supporting evidence, and give an example; (e) briefly rephrase what you have said, to ensure everyone understands each stage in your talk; and (f) restate your main points towards the end of your talk so that they lead directly to your conclusions.

Similarly, in a presentation it might be appropriate to: (a) show your understanding of the present position, so far as your audience is concerned; (b) refer to the reason why changes are being considered; (c) discuss possible courses of action and make a recommendation; and (d) repeat your main points.

Afterwards, if necessary, provide a handout that repeats your main points and provides more detail than would have been appropriate in your talk or presentation.

If you have agreed to talk on a particular subject, keep to your terms of reference. Decide on a limited number of main points that you must make. Arrange these in an appropriate order (as a topic outline for your talk); then check that they are all essential in relation to your aim.

You will find it helpful to make a note of each main point on an index card or at the top of a blank sheet of paper – with any essential supporting details or evidence summarised below each heading (and notes in the margin to indicate where you should be at different times after the start). The number of points you can make in the time available, and the amount of supporting detail required, will depend upon your audience. What prior knowledge, if any, can you assume to be shared by everyone likely to be present?

Plan any demonstration that will reinforce your words and add interest. Consider if any visual aids are required to support your words. At least, decide which words are most important (these are your main headings) and which words may be new to some members of your audience, so that you will remember to spell out and, if necessary, to define them during your talk.

If you use visual aids in your talk you will be able to say less – but may convey more information – and pictures may be remembered even if much of what you say is forgotten. Visual aids add interest and provide a change for the audience – from listening to seeing. They capture attention, and so should be used only for important points. They also help to hold attention if, for example, they are used in sequence as you develop an argument. They should complement your words, enabling you to provide essential evidence (for example, in a table or graph) that could not be conveyed adequately with words alone.

If you use any aid that is to occupy a few minutes (for example, using closed-circuit television, a film or a video) it is usually best to include this so that it provides a break about half-way through your talk. Your first words will then provide an introduction, and your last words your conclusions.

Prepare any necessary stores, equipment, handouts or visual aids (see pages 168–9); and decide exactly when you will use them – so that they support your spoken words and are not a distraction when you are trying to interest your audience in something else. For example, do not provide handouts before a talk unless they are needed during the talk.

Obtaining a response. Try to make your talk interesting. This will depend upon: (a) your knowledge of the subject and your ability to select just what is relevant to this talk; (b) showing that what you have to say is relevant to the needs of this audience – that it follows on from their existing interests or that it will help them in some other way; (c) ensuring variety and simplicity in your presentation; (d) using audio and/or visual aids and other demonstrations so that people see and touch, as appropriate, and hear other relevant sounds as well as listening to your voice; and (e) avoiding distractions. People will listen most carefully, and will remember best what you say, in the first fifteen minutes of a talk; and thirty minutes with one person talking is enough for any listener. A well-planned thirty-minute talk may comprise a brief introduction (5 minutes), your main points (10 minutes), elaboration and visual aids (8 minutes), your conclusions (2 minutes), and questions (5 minutes).

Try to anticipate questions that are likely to be asked, so that you are prepared to give concise answers or able to say where further information is to be found.

Write your talk in full, with appropriate headings for each main point and marginal notes to remind you when to use your visual aids. If you plan to talk for 25 minutes, you should write about 2500 words (about 10 sheets of A4 paper typed double spaced, using one side only).

Read your script aloud to yourself, pausing where necessary (for example, where you plan to use a visual aid) to check how much time you need to make each point, that you can complete the talk in the time allowed, and that you have not written anything you would not say.

Rehearse your talk, referring only to brief notes (one sheet of paper or a cue card with a heading for each topic and a few key words, similar to the outline prepared when you were deciding what to say) to ensure you can finish on time. Ask colleagues to listen and let you have any comments, questions or suggestions. Then, during your talk, if you can, refer only to these brief notes – so that for most of the time you are looking at your audience.

Many people have favourite words . . . *sort of, like, er, I mean* . . . which they repeat so often that the listeners' attention is distracted from the important words. Unwanted words and phrases such as these, which give the speaker time for thought, may be a sign of inadequate preparation or nervousness. Other expressions – *you see, you know, all right*, and *if you follow me* – are attempts at confirmation. You may find it helpful to record your talk, to check that it sounds well, that each of your main points is made effectively, that you have made proper connections, and that you do not have particular words or expressions that you over use.

Find out the size of the room to be used for your talk, so that you can ensure that everyone present will be able to see any demonstration and read the words on your visual aids. Check that any equipment you need is available, that you know how to use it, and that it is in working order.

Preparing visual aids

One advantage of using a blackboard, whiteboard or flip chart is that you have to prepare effective visual aids quickly at the most appropriate times during your talk – so each one must be clear and simple, and you cannot prepare more than you can complete and your audience assimilate in the time available.

However, most speakers like to prepare their visual aids before a talk. This saves time during the talk, but a common result is that a speaker uses too many visual aids and says too much – in an attempt to present more information than should be included in a talk.

Do not prepare too many visual aids. You may, for example, decide to use one visual aid to reinforce each of your main points. A talk or presentation is an opportunity for people to see and hear a speaker, to consider what is said, and to ask questions. It should not be just a slide show.

Do not use a visual aid if it includes too many words, too much detail, or anything that is not relevant to your talk. Use one visual aid to convey one message and make that message brief, clear and simple – so that it can be understood quickly. A useful guide is to include no more than 8 lines of writing and no more than 8 words per line, but not more than 32 words in total. The fewer the words used, the better is the visual aid. A labelled diagram or drawing, or a cartoon, is effective because it has a picture as well as words. Whereas people can look and listen together, they read at different speeds. Listening unites them, reading sets them apart. So, as a rule, prefer a diagram, drawing or photograph to a written message.

Do not prepare a table that includes too many numbers or has words or numbers that are too small for people to see them clearly. Tables and illustrations from books are likely to include more detail than is acceptable in a visual aid, unless they were prepared with both uses in mind.

Tables that are to be photographed and made into slides can be prepared with a word processor. The title should be one line of up to 7 words, and there should be up to 4 columns and up to 8 horizontal rows. Print in double spacing, with 12 point font size in a rectangle 12 cm x 8 cm, or with 10 point in a rectangle 10 cm x 6.5 cm. Alternatively, with appropriate software, visual aids prepared using a computer can be stored electronically on a disk (see *Preparing presentations*, page 192)

However, do not use special effects, fancy lettering or elaborate backgrounds just because they are available in a computer program. Each visual aid should convey your message as clearly and simply as you can. Use white or yellow lines and lettering on a black, dark blue or mid-green background, or use black or dark colours on a white background. Also note that if your visual aids are to be used in a handout, or publication, prepared with a monochrome printer, black on a white background is best.

Using a blackboard, whiteboard, or flip chart

1   Spell any words that may be new to some people in your audience, in large, clear, capital letters. If necessary, define them.

2   With a blackboard or whiteboard, prepare clear, simple diagrams quickly during your talk (but plan them before your talk so that you are in control).

3   With a flip chart, key words, definitions and simple diagrams (or blank sheets) can be arranged in order before your talk, and then displayed as they are required. If you do this, leave a blank sheet after each visual aid so that your audience cannot see the next one until it is

needed. Also, if you prepare your charts in advance, you may find it helpful to prepare them in reverse order, so that you can flip them forwards easily during your talk.

One advantage of preparing your visual aids before a talk is that you can check that they can be seen by everyone in the room and that the smallest numbers and letters are easy to read from the back row. However, if you prefer your audience to see how your argument develops or how a diagram is constructed, you may find it helpful to prepare your visual aids in advance but in soft pencil, so that the words or the lines of your diagram will not be seen until you go over them using a broad-tip water-colour marker.

4   When using a board or chart, try not to obscure anyone's view. If you are right handed, stand to the left while you are writing or drawing. Afterwards, point with your left hand so that you can face your audience and everyone can see the visual aid.

5   If you turn away from your audience, to write or draw, stop talking.

6   Because people can observe as you write or draw, you may not need to allow more time for them to study your work before proceeding with your talk.

7   Always have contingency plans for use of a board or flip chart, in case for any reason you are unable to use your other visual aids.

Using an overhead projector

As with any other equipment, you will find an overhead projector most useful if you have considered how best to use it:

1   Check, from the back of the room in which you are to give your talk, that any diagrams and tables are clear and that you have not included too much detail or anything irrelevant.

2   Place your first transparency on the projector immediately before you start to speak, so that when you are ready to use it you have only to switch on the light.

3   If you write during your talk, make sure that the lines are distinct and that the words are legible. Spell, in large capital letters, any words that may be new to some people in your audience. If necessary, define them.

4   Use a pointer so that you can look at your audience even when you are pointing at the screen. When you look at the screen, your audience will look at the screen, and if you then look at your audience you will see that they are looking at the screen or, when you speak, that you have their attention. You are advised not to point at the transparency,

because you cannot look down and up at the same time. However, if you do point at the transparency, do not point with your finger – which would obscure too much of your visual aid.

5 You may find it helpful to cover part of a table or diagram with a card, so that you can display just the parts required at the time. Alternatively, you can build up a diagram in stages by superimposing transparencies.

6 When you write or draw, stop talking. Then give people time to study any diagram quietly before you explain or continue with your talk.

7 It is usually most convenient to use a separate transparency for each visual aid, and to remove it as soon as it has served your purpose. However, some overhead projectors have a roll of acetate film on which, for example, you can: (a) write during your talk, and then turn the roll to remove your writing; or (b) prepare a sequence of drawings or diagrams, linked by numbers or arrows, so that you can roll them into view, like a flow diagram, to show successive stages in a process. Alternatively: (c) you can fix a transparency below the roll before your talk, to display a drawing, diagram or outline map, so that labelling, symbols or additional artwork can be superimposed by drawing or writing on the roll during the talk – before you turn the roll on so that you can use the same diagram to help you make different points at successive stages of your talk.

8 Try to ensure that people are looking at what you wish them to see: give them time to look at each transparency, but remove it as soon as you are ready to move on.

9 Look at your audience when you speak.

10 Stand away from the projector – next to your notes. This may be particularly important if you are using a microphone.

Using slides

If you decide to use slides, consider when best to use them. If it is necessary to switch off the room lights to use a slide projector, it is disturbing to everyone if you switch them off and on repeatedly; but if the lights are off all the time you cannot look at your audience and they cannot make notes. Try to show the slides in one batch. Talk first and then show your slides; or use the slides to provide a break in a long talk or to separate the body of your talk from the summary and conclusions. What is best on one occasion may be inappropriate on another.

1 Sit at the back of a room, similar in size to the room in which you will be talking, and check your visual aids for clarity. Do not show a list,

diagram or table if it has too many words or numbers – or if any are so small that some people cannot read them.

2   Arrange the slides in the same order as in your notes.

3   If you are actually using slides (not images stored electronically), check that each illustration or table is the right way up and the right way round when projected on to the screen.

4   Give the audience time to look at each slide, then say what you want them to note.

5   If you use a pointer, keep it still on the point you want people to observe. Then put it down.

6   As with other visual aids, remove each slide as soon as it has served its purpose.

Delivering your talk

Complete the final arrangements shortly before your talk. Make sure you know how to use any switches or projection equipment, or brief the projectionist. If possible, ensure that the room is warm enough but well ventilated, and that there are no distracting noises. Stand where everyone can see you, but avoid distracting mannerisms such as hand movements that convey no meaning, swinging or banging a pointer, or constantly walking to and fro as if on sentry duty. This is not to say that you should not move your hands: some speakers use carefully considered gestures to good effect.

Some speakers use a joke to put people at ease; but it may be difficult to find a new joke that matches the interests of your audience and is appropriate for the occasion – and you will not get off to a good start if people feel obliged to laugh.

In preparing a talk, most speakers write down exactly what they plan to say and then practise to ensure they can complete the talk comfortably in the time available. However, if possible, do not read your talk; and do not simply read aloud the words displayed on your visual aids – which people can read for themselves.

Make sure that everyone knows who you are! If you are the first speaker in a joint presentation, you should also welcome the audience – if they have been invited, introduce the other presenters and explain their role, and say briefly how the whole presentation is to be made.

Speak so that everyone can hear every word, but do not use a microphone unless poor acoustics make this necessary. Try not to speak in a monotone. Show your enthusiasm for the subject and your interest in everyone present. Look around your audience so that you can capture and maintain attention, and everyone can see your facial expressions.

Say what you are going to talk about. Remind the audience how this follows on from what they already know. Give the reason for your talk. Define your aim. This is your opportunity, in your introduction, to capture attention and promote a desire to listen.

Use your notes as reminders, but look at your audience while you are speaking. Maintain eye contact so that you are aware of those people who understand and of those who require further explanation. Stop talking whenever you face away from your audience (for example, to write a word on a blackboard or whiteboard).

Get to the point quickly at the start of each aspect of your talk. Pause briefly after each main point has been made. This pause will emphasise the point, let everyone know that it is time to start thinking about something else, and give you time to refer to your notes.

To ensure that you keep their attention, it is a good idea to give your audience something to do. For example, you may ask a question of your audience from time to time – to make them consider something relevant to your next point. Then pause briefly to give everyone time to think before you *either* answer the question yourself *or* invite one person, by name, to attempt an answer. If you write a word on a blackboard or whiteboard, or use some other visual aid to reinforce a main point, so that there is something for your audience to see as well as something to hear, remember to allow them enough time to study it without the distracting sound of your voice.

If you use a blackboard or whiteboard, keep it clean. If you use a flip chart, turn to a clean sheet as soon as you have made your point. Remove any visual aid as soon as you have finished with it. Do not allow people to continue looking at one thing while you are trying to interest them in something else.

Bring your talk to an effective conclusion. Summarise each of your main points and state clearly what conclusions you draw. Say why they may be important to your audience.

Allow time for questions. The questions, and your answers, may add relevant information or ideas, or help to prevent misunderstandings. Remain at the front of the room facing your audience. Make a note of any question, and repeat it to make sure that everyone knows exactly what the question is. Then keep your answer short, clear and to the point.

Finish on time. Nobody will mind if your talk ends a few minutes early, but do not speak for too long. In a well-organised meeting the chairman will ensure that you start on time. Then you must time each stage in your talk so that you say what you intended to say, in the way you planned to say it, and finish on time – after answering a few questions. Leave people to reflect on your words. If you talk for too long they may remember only that you did not know when to stop.

Appendix A   Punctuation

Some people suggest that mistakes in grammar and punctuation do not matter if the writer's meaning is clear; but if the English is poor, the meaning is unlikely to be clear. For example:

This latest outbreak of violence has not surprisingly received the condemnation of politicians of all parties.

To make clear whether or not the rioting has been condemned, commas are needed in the above sentence – after *surprisingly* and either before or after *not.*

This latest outbreak of violence has, not surprisingly, received . . .

This latest outbreak of violence has not, surprisingly, received . . .

However, it is best to use no more words, and no more punctuation marks, than are needed to make your meaning clear. So, to make the meaning clearer, without punctuation, it would be better to write:

either   This latest outbreak of violence has been condemned by . . .

or   This latest outbreak of violence has not been condemned by . . .

Using punctuation marks to make your meaning clear

In writing, punctuation marks indicate pauses – and other characteristics of speech – which help to make your meaning clear. For example:

The Prime Minister said, 'The Leader of the Opposition is a fool.'

'The Prime Minister', said the Leader of the Opposition, 'is a fool.'

The meaning of the first of these sentences is the opposite of that of the second, but the words, and the order of words, are identical. Only the punctuation marks differ.

Table A1.1 Parts of speech: classifying words

Parts of speech	The work words do in a sentence
Verbs	Words used to indicate action: what is done, or what was done, or what is said to be. Earth *rotates* around the sun.
Nouns	Names. The *sun* is a *star*.
Pronouns	Words used instead of nouns or so that nouns need not be repeated. The sun is a star. *It* is at the centre of our solar system.
Adjectives	Words that describe or qualify nouns or pronouns. The sun is a *small* star.
Adverbs	Words that modify verbs, adjectives and other adverbs. The sun is a *fairly* small star.
Prepositions	Each preposition governs, and marks the relation between, a noun or pronoun and some other word in the sentence. Earth rotates *around* the sun.
Conjunctions	Words used to join the parts of a sentence, or to make two sentences into one. The sun is at the centre of our solar system, *but* it is just one fairly small star near the edge of one of the very many galaxies in the universe.

If you have difficulty with punctuation, you will find it easiest to write in short sentences. For example:

A sentence begins with a capital letter. It includes a verb. It ends with a full stop. It expresses a whole thought, or a few closely connected thoughts. It therefore makes sense by itself.

Each of the last five sentences expresses one thought, telling the reader one thing about a sentence, but if you were to write only in short sentences, your reader would have no sooner started reading each one than it would be time to stop. For people who read well, therefore, a whole document written in short sentences would be hard reading – not easy reading.

Using punctuation marks to ensure the smooth flow of language

The thoughts expressed in the five short sentences can be expressed in two:

A sentence starts with a capital letter, includes a verb, and ends with a full stop. Because it expresses a whole thought or a few closely connected thoughts, it makes sense by itself.

In different sentences you may use the same words to express different thoughts.

You can help. Can you help?

Conversely, in different sentences you may use different words to express the same thought:

> Come! You come. Come here, you!

Using conjunctions to contribute to the smooth flow of language

Conjunctions (for example, *and, but, for, when, which, because*) can be used to join parts of a sentence or to make two sentences into one (see Table A1.1). They link closely related thoughts, give continuity to your writing, and so help your readers along. However, use each conjunction intelligently, and if possible not more than once in a sentence. Remember, also, that some conjunctions must be used in pairs: *both* is always followed by *and*; *either* by *or*; *neither* by *nor*; and *not only* by *but also*.

Using capital (upper case) letters

Capital initial letters are used for the first word in a sentence or heading, for most words in the titles of publications (see page 196), for proper nouns (for example, trade names, see page 64), for interjections, and for most acronyms and titles (see page 65):

> Our church is St Ann's Church.

Whole words in chapter headings, and in the section headings of a report, may be written in capitals. Otherwise, capital letters are rarely used for whole words – and initial capital letters are no longer used to emphasise words within a sentence that are not proper nouns (see page 71).

In handwriting a clear distinction should be made between upper and lower case letters; and (except possibly in a signature) capitals should not be used as an embellishment.

Punctuation marks that end a sentence

If you find punctuation difficult, begin by mastering the use of the full stop and keep your sentences short and to the point.

Full stop, exclamation mark and question mark

The end of a sentence (or interjection) is indicated by a full stop, exclamation mark or question mark.

> You must go. Go! Must you go?

Remember that a question mark is used only after a direct question:
    Please explain.
    Could you explain, please?
    I should appreciate an explanation.
    I wonder if I should ask for an explanation.

Punctuation marks used within a sentence

The punctuation marks used to separate parts of a sentence make the reader pause for a shorter time than does a full stop. The more you read and write, the more you will come to appreciate their value in helping you to communicate your thoughts precisely.

Comma

Items in a list may be separated by commas, as in the next sentence. To write clear, concise and easily read prose we use commas, semicolons, colons, dashes and parentheses. In such a list the comma is essential before the final *and* only if it contributes to clarity.

A comma may also be used to separate the parts (or clauses) in a sentence. The word clause comes from the Latin word *claudere*, to close, and within a sentence commas may be needed to separate (close off) one thought or statement from the next.

A sentence comprising one clause, expressing one thought, is called a simple sentence. It makes one statement:
    Each word should contribute to the sentence.
    Each sentence should contribute to the paragraph.
    Each paragraph should contribute to the composition.
    Nothing should be superfluous.

However, a sentence may comprise more than one clause – expressing more than one thought. A comma or a conjunction, or both, may then be inserted between the separate statements (clauses):
    Each word should contribute to the sentence, each sentence to the paragraph, and each paragraph to the composition. Nothing should be superfluous.

Note that in this example, at the beginning of the second clause the conjunction (*and*) is understood: there is no need to write it. Similarly, in each clause there is a verb, but in the second and third clauses this verb (*contribute*) is understood.

Use commas to mark separate clauses if they make for easy reading and help you to convey your thoughts. A commenting clause should be enclosed by commas; a defining clause should not be.

Nurses, who work on Sundays, are . . .
Nurses who work on Sundays are . . .

Note the difference in meaning. The first sentence implies that all nurses work on Sundays. The second sentence identifies or defines which nurses are referred to: those who do work on Sundays.

A comma is used either after, or before and after, some adverbs, for emphasis, as in the following examples.

However, . . .
There are, however, . . .
Therefore, . . .
Note, therefore, that . . .

Do not add commas at random because you feel that a sentence is too long to be without punctuation marks. Either put the comma in the right place, to convey your meaning, or write the sentence so that your meaning is conveyed clearly without the comma.

You will be informed, if you send a stamped addressed envelope, after the meeting.
You will be informed, if you send a stamped addressed envelope after the meeting.
If you send a stamped addressed envelope you will be informed after the meeting.

Note that the first and third sentences convey the same message; one with commas and the other without.

Parentheses and dashes

Parentheses (curved brackets) are always used in pairs, and dashes may be used – in pairs – when an aside is added to a sentence. So if you removed the asides from the last sentence, you would be left with a complete sentence. Each aside is said to be in parenthesis. Use parentheses when you wish to insert a cross-reference (see page 176), an example (see page 173), or an explanation (as in the first sentence of this paragraph). Use dashes to give prominence to an insertion (as in the first sentence of this paragraph). But note that commas could be used instead of dashes, as in this sentence, if you wished to give less prominence to an aside. One dash can be used if an aside is added at the end of a sentence – as in this sentence. See also square brackets, page 128 and page 180.

Colon

A colon may be used to introduce a list (as on page 12 or a quotation (as on page 11); and may also be used, in place of a full stop, either (a) between two statements of equal weight (as on page 109) or (b) between two statements if the second is an explanation or elaboration of the first (as on page 48).

Semicolon

The full stop (or period), the colon, the semicolon, the dash, the comma, and bracket are all punctuation marks, points or stops. They all indicate pauses. The full stop gives the longest and most impressive pause. The colon gives a shorter pause. Use of the semicolon, which gives a shorter pause than a colon but a longer pause than a comma, may contribute to clarity (for examples, see pages 45, 50, 84, 93 and 126).

Other marks

Apostrophe

The apostrophe is the mark that causes the greatest or most obvious difficulty for many educated English-speaking people. If you are not sure about its use, first note that an apostrophe is *never* used in forming the plural: apple becomes apples; criterion, criteria; datum, data; gateau, gateaux; lady, ladies; man, men; mouse, mice; phenomenon, phenomena; and wife, wives. Then note that if, in writing, you avoid colloquial language (see page 186), in which an apostrophe is used to mark a contraction (for example, *can't* for *cannot, don't* for *do not, it's* for *it is* or *it has, that's* for *that is, there's* for *there is, they're* for *they are, who's* for *who is,* and *won't* for *will not*), you will use an apostrophe only when you wish to indicate that someone or something belongs to someone or something (see page 184).

An s is added to many nouns (names of things, see Table A1.1) to make them plural: book becomes books; but man becomes men. To indicate ownership either an apostrophe s ('s) is added to a word (book's and men's) or just an apostrophe is added (books'). For example: the cat's dinner (the dinner of the cat); the cats' dinner (the dinner of the cats); the man's books (the books belonging to the man); the men's books (the books belonging to the men); the books' covers (the covers of the books); the book's cover (the cover of the book).

Write Dr Smith's office (the office of Dr Smith) but *either* Dr Jones' office *or* Dr Jones's office is acceptable (for the office of Dr Jones).

Note that 1990's music (apostrophe before the s) is the music of 1990, and 1990s' music (apostrophe after the *s*) is the music of the 1990s (the ten years from 1990 to 1999), no apostrophe being required when simply forming the plural.

Do not add an apostrophe and *s*, to indicate ownership, to a word that in itself indicates ownership (a possessive adjective or a possessive pronoun – see list on page 184).

Quotation marks

You may use quotation marks when you quote someone else's words exactly (as on pages 13 and 74), or you may sign-post extracts by a footnote in a table (as on page 87) or by indentation in the text and an acknowledgement (as on page 11) – including quotation marks only if they are part of the extract (as on page 63).

When quoting someone else's work, the part quoted must be complete – including every word and every punctuation mark. Any gaps in the quotation should be indicated (by three dots, as on page 74) and any words you insert must be in square brackets (as on page 84). The source of each quotation should normally be acknowledged (see page 13), unless you have some good reason for not doing so (for example, see pages 76–7).

The use of quotation marks (inverted commas) to indicate that a word or phrase is not to be understood in its usual sense is to be avoided, because the intended sense may not be clear to the reader. Instead, choose words that convey your meaning precisely (see page 55).

The titles of books, plays and poems should not be in quotation marks, as is sometimes recommended. Instead, in handwriting they should be underlined and in word processing or print they should be in italics. Use underlining or italics to help you distinguish, for example, between David Copperfield (the name of a character in a book) and *David Copperfield* (the title of a book).

Improve your writing

The best way to appreciate the usefulness of different punctuation marks is to study one or two pages of any book or article that interests you. Consider why the author has used each punctuation mark. You can repeat this exercise with as many compositions as you choose to study. In writing clear and simple English you can manage without semicolons and colons, but as you begin to appreciate their value you will want to use them.

Appendix B   Spelling

Mistakes in spelling, as with mistakes in punctuation and grammar, reduce an educated reader's confidence in a writer. They also distract readers, taking their attention away from the writer's message. Spelling correctly, therefore, is part of efficient communication.

Some reasons for poor spelling

Some words are not spelt as they are pronounced: for example, answer (anser), gauge (gage), island (iland), mortgage (morgage), psychology (sycology), rough (ruff), sugar (shugar) and tongue (tung). You cannot, therefore, spell all words as you pronounce them. This is one problem for people who find spelling difficult.

However, those who speak badly are likely to find that incorrect pronunciation does lead to incorrect spelling. In lazy speech secretary becomes secatray; environment, enviroment; police, pleece; computer, compu'er; and so on. If you know that you speak and spell badly, take more care over your speech.

Unfortunately, the speech of teachers and that of announcers on radio and television does not necessarily provide a reliable guide to pronunciation. Consult a dictionary, therefore, if you are unsure of the pronunciation or spelling of a word. And, when you consult a dictionary to see how a word is spelt, check the pronunciation at the same time. Knowing how to pronounce the word correctly, you may have no further difficulty in spelling it correctly.

If you do not read very much, you give yourself few opportunities for increasing your vocabulary (see Chapter 6) and for seeing words spelt correctly. Reading good prose will help you in these and other ways.

Some rules to remember

The best way to improve your spelling is to consult a dictionary when necessary and to memorise the correct spelling of any words that you have found difficult. However, learning the following rules – one at a time – will also help.

1   When *ie* or *ei* are pronounced *ee*, the *i* comes before the *e* except after *c* (as in believe and receive).
Exceptions to this rule are seize and species.
In eight, either, foreign, freight, reign, their, weight and weir the *ei* is not pronounced *ee*, so the *i* does not come before the *e*.

2   When words ending in *fer* are made longer (for example when refer is used in making the longer words reference and referred) the *r* is not doubled if, in pronouncing the longer word, you stress the first syllable (as in *re*ference), but it is doubled if you stress the second syllable (as in re*ferred*). A syllable is a unit of pronunciation which forms a word or part of a word.

	First stress	*Second stress*
defer	*de*ference	de*ferred*, de*ferring*
differ	*differ*ed, *differ*ence, *differ*ing	
infer	*in*ference	in*ferred*, in*ferring*
offer	*offer*ed, *offer*ing	
refer	*re*feree, *re*ference	re*ferred*. re*ferring*
suffer	*suffer*ed, *suffer*ing, *suffer*ance	
transfer	*trans*ference	trans*ferred*, trans*ferring*

3   With verbs of more than one syllable that end with a single vowel (*a*, *e*, *i*, *o* or *u*) followed by a single consonant (a letter that is not a vowel), in forming the past tense or a present or past participle double the consonant if the last syllable is stressed:

	First stress	*Second stress*
benefit	benefited, benefiting	
bias	biased	
control		controlled, controlling
excel		excelled, excelling
focus	focused, focusing	
parallel	paralleled	
refer		referred, referring

There are exceptions to this rule, including funnel (funnelled), model (modelled), panel (panelled), rival (rivalled), travel (travelled) and tunnel (tunnelled).

4 With verbs of one syllable that end with a single vowel followed by a single consonant, double the consonant before adding *ing*:

run	running
sag	sagging
swim	swimming
whip	whipping

But if a verb of one syllable does not end in a single vowel followed by a single consonant, simply add *ing*:

daub	daubing
deal	dealing
feel	feeling
help	helping
sink	sinking
watch	watching

5 When verbs ending in *e* are made into words ending in *ing* the *e* is lost:

bite	biting
come	coming
make	making
trouble	troubling
write	writing

But there are exceptions:

agree	agreeing (to keep the *ee* sound)
dye	dyeing (colouring)
flee	fleeing (to keep the *ee* sound)
hoe	hoeing
singe	singeing (to keep the soft *g*)

And with some verbs the *ie* ending is replaced by *y*:

die	dying
lie	lying

6   If an adjective (see Table A1.1) ends in *l*, the corresponding adverb (which answers the question How?) ends in *lly*:

     beautiful   beautifully
     faithful   faithfully
     hopeful   hopefully
     peaceful   peacefully
     spiteful   spitefully

7   Some adjectives that end in *y* have corresponding adverbs and nouns in which the *y* is replaced by an *i*:

     busy   busily   business
     merry   merrily   merriment

Many people have difficulty in spelling some words correctly because they are unable to distinguish between there and their, its and it's, whose and who's, book and book's, books and books'. If you cannot decide which to use, avoid colloquial language (see page 179) and learn how to indicate ownership.

Their and theirs are used to indicate that something belongs to some people or to some thing. Remember this rule: *e* in her, *i* in his, *e* and *i* in *their* and *theirs*, to indicate possession.

There is used with the verb *to be*. Remember: there is, there are, there was, there were t h e r e spells there. This spelling is also used for a place:

Is anyone there? There is their house, over there.

My, his, her, its, our, your and their are possessive adjectives: my book, her eyes, its leaves, and their house. Mine, his, hers, its, ours, yours and theirs are possessive pronouns:

This book is mine; this is yours, and these are theirs.

To distinguish its (possessive) from it's (colloquial: a contraction), remember first that it's means *either* it is *or* it has, and second that colloquial language should not be used in scientific writing. Instead, use standard English or standard American (see page 186). See also *The apostrophe*, page 179).

Improve your writing

Keep a good dictionary on your bookshelf

Always have a good dictionary (see page 65) available for reference when you are thinking, reading or writing. Do not get into the habit of using another word when you are unsure of the spelling of the most appropriate word. Instead, always refer to a dictionary so that you can use the word that best conveys your meaning. Make a note, from your dictionary, of the correct spelling of any words you spell incorrectly (as indicated, for example, by a spell checker on your computer) so that you can memorise the correct spelling.

Spelling test

Ask someone to test your spelling of these words:

absence, accelerate, accessible, accidentally, accommodate, achieve, acquaint, address, advertisement, altogether, analogous, ancillary, apparent, attendance, audience, auxiliary

beautiful, beginning, benefited, bureaucracy, business

calendar, census, cereal, certain, competence, conscience, conscientious, conscious, consensus, commitment, committee, correspondence, criticism

decision, definite, desiccated, desperate, develop, disappear, disappoint

embarrass, environment, eradicate, especially, exaggerate, existence

faithfully, fascinate, February, forty, fourth, fulfil, fulfilled

gauge, government, grammar, guarantee

harassment, harmful, height, hierarchy, humorous,

idiosyncrasy, incidentally, independent, irradiate

liaison, library, loose, lose, lying

maintenance, management, misspell, millennium, minuscule, minutes

necessary, noticeably

occasion, occurrence, omit, omitted

parallel, parliament, planning, personnel, possess, precede, privilege, procedure, proceed, profession, pronunciation, publicly, pursued quiet, quite

receipt, receive, recommend, relevant, restaurant, rhythm

scissors, secretary, seize, separate, severely, siege, sincerely, successful, supersede, surprising, syllable

unnecessarily, until

Wednesday, wholly

yield

Take an interest in the study of the origins of words (etymology)

Knowing the origin of a word may help you to understand its spelling. For example, the word *separate* is derived from a Latin word *separare* (to separate or divide); so is another English word, *pare*, meaning to cut one's nails or to peel potatoes; but *desperate*, from the Latin *sperare* (to hope), means without hope.

Write in standard English

Scientists should write in standard English (or standard American) and should avoid colloquial English and slang.

> *Standard English*: the language used by educated English people.
> *Colloquial English*: the English used between close friends, including such contractions as don't and won't.
> *Slang*: highly colloquial language including new words or words used in a special sense which might not be understood by educated English people.

Partridge (1965) gives, as an example of the difference: man (standard), chap (colloquial) and bloke, cove, cully, guy, stiff, or bozo (slang).

Appendix C
Computer appreciation

Using your computer

People who cannot touch-type are handicapped when using a computer keyboard. Many hand write at least the first draft of anything other than a very short composition so that they can work fast enough to allow their thoughts and their written words to flow. Then they spend more time than should be necessary word processing later drafts.

So, if you cannot touch-type you are advised to learn, preferably before using a computer for word processing. You could learn from a book that includes basic instructions and graded exercises, or attend a class on keyboard skills, or buy a computer program that provides on-screen instruction. With regular and frequent practice, you should soon be typing faster than you can write.

Word processing

With a personal computer containing appropriate software you can produce pages of text, including tables and illustrations, with a print quality similar to that of a book. However, you are advised not to justify right-hand margins, and not to use bold, italics or underlining to emphasise words in the text of a document (except that italic print is used for the words *either* and *or* if it is necessary to emphasise an important distinction). Capitals, bold print and italics can be used for different grades of headings (see page 136), and most headings should be given a line to themselves – for emphasis – so there is no need to underline them. Italics, or underlining, can also be used for words that in a handwritten composition should be underlined (see page 154).

If some users think of a word processor as a tool that eliminates the need for thinking and planning before writing, and for care in writing, because it is easy to correct and revise their work later, they are wrong. A computer has a memory but no intelligence. It is a tool that can make writing easier, but the writer still has to do the thinking at each stage in composition.

When working on a screen, as in writing with a pen, you must: (1) make notes as you think about what is required; (2) rearrange your notes below appropriate headings as you prepare a topic outline for your composition; (3) choose and arrange words carefully as you write, to ensure you express your thoughts clearly and simply; and then (4) check, correct and if necessary revise your work (see pages 41–6). As a result, there should not be much wrong with your first draft. If there is, nothing you can do in checking and revising can compensate for your not having considered the needs of the reader, or for not devoting enough time to thinking and to planning, before starting to write. This is true whether your composition is handwritten or wordprocessed.

Because with a word processor it is so easy to make additions and deletions, to cut and paste, and to copy, great care is needed in checking a document to ensure that it reads well, with no words missing and no words, sentences or paragraphs duplicated or out of place (see *Revising*, page 46, and *Checking your typescript*, page 155)).

Use the spell checker on your computer. It will help you to correct typing errors and spelling mistakes, and so to improve your spelling. However, although a spell checker ensures that each word used is spelt correctly (in American English or British English) it does not ensure that it is the right word (see pages 55–9). For example, does the spelling and grammar checker on your computer draw attention to any errors when you type the following sentences?

I advice you to consider the following advise.
There's too mistakes in the last sentence.

There are, in fact, two mistakes in each of these sentences. They should read:

I advise you to consider the following advice.
There're two mistakes . . . (There's means There is)

However, in business it is best to avoid colloquial language by writing:
There are two mistakes . . .

Also, do not allow a spell checker to spell-check and change, automatically, specialist terms, abbreviations, acronyms, or proper names (of people and places) unless these are correct in your computer's spell-check dictionary. It would be embarrassing, for example, if the computer changed Mr Charlton's surname to Charlotte or worse to Charlatan – and you did not notice the mistake when checking the document.

If you are a student, you are advised not to use a word processor for all your assessed course work. If you do, you may find it very difficult to think, plan, write quickly and check your work in examinations – when you have to use handwriting and do your best work. In course work you can spend more time on thinking and planning than you could spare in an examination, but at least the first draft of your answer should be handwritten in about the time that would be available in an examination – when you would not be able to use a word processor.

Indeed, in course work students who can prepare a neat handwritten first draft that is legible and well presented – so that it does not need to be revised – should not be required – as they are on many courses – to waste their time word processing a second draft just to change their handwriting into print.

Both students and their assessors should accept that a composition can be well presented without its being *Control*, word processed (see page 81). Whether or not you are a student – you must develop the ability to write quickly and to get things right the first time, even if to write quickly your composition has to be handwritten.

Looking after your documents

1 Information obtained via the Internet, including attachments to incoming e-mail messages, might be contaminated with viruses, and should be checked before opening.

2 Before using a computer, therefore, ensure that it has up-to-date virus-detecting and virus-removing software installed.

3 Before using a disk for the first time, ensure that it is checked for viruses with an up-to-date virus checker.

4 When producing a new document, use a new disk and back-up disk for just that document.

5 Save (or file) your work frequently, as you plan, write, correct, or revise a document, so that if anything is lost (for example, as a result of a power failure) you do not lose much of the document and can try to do the work again quickly while the information and ideas are still fresh in your mind.

6 Save your work before you try any new commands if there is any possibility that you may lose or inadvertently alter part or all of the document, so that you can quit (that is, leave the document in its original state) and try again.

7 Your floppy disks may go wrong, as may the hard disk of your personal computer, causing you to lose all your work at any time. So ensure that all data stored in a computer are backed up with a frequency that

reflects their value and importance. Take a local copy immediately after data have been entered from memory, or from an enquiry or investigation. Each day, when working on a document, make a new copy using a different file name (for example, the year, month and day). If you are working on a document for several days, or for several weeks, take daily, weekly and monthly back-ups on separate disks. Bear in mind that disks are inexpensive, whereas your time spent in re-entering lost information – if this were possible – would cost much more and would interfere with your other work.

8 Label your disks consecutively (for example, with your initials and a number: ABC001, ABC002, etc.) and maintain a log of your disks in a small hardback notebook. Record what each disk contains, and for back-up disks record the type of back-up (daily, weekly, or monthly).

9 When a document is complete, copy it into your master archive disk, and back-up archive disk, in case you need copies later, or need to update it, or include parts in another document.

10 Reformat your document disk ready for your next document.

11 Do not carry all your disks with you at one time. Keep your master archive and master back-up disks in separate places, so that if one is lost or damaged you still have the other.

Looking after yourself

1 Sit comfortably at your computer. Adjust your chair so that you are close to the desk, with your elbows level with the computer keyboard, your feet resting flat on the floor or on a foot rest, and your back upright. When using a mouse, rest your arm on the desk and move your hand by moving the elbow rather than the wrist. If you touch-type, you could try using a contoured keyboard.

2 Adjust the height of the visual display unit, if necessary, so that your eyes are level with the top of the screen and 30 to 60 cm from the screen.

3 Ensure the screen is clean and free from glare (for example, from a lamp or window) and that the keyboard and adjacent work surface are sufficiently illuminated – but have a matt surface that does not reflect light.

4 If necessary, adjust the brightness and contrast controls on your visual display unit, so that the background is no brighter than is necessary for you to see the words clearly.

5 If you cannot touch-type you will find it tiring to be constantly looking down at the keyboard, and at your handwritten draft, and then up at the screen. But if you can touch-type, you will not need to look at the keyboard when copy typing and may find it helpful to use a document holder to hold your papers adjacent to the screen.

6   Do not allow the use of a computer to become an end in itself. A computer helps you to do many things, some of which would not otherwise be possible (for example, in recording, processing, storing and retrieving information); but in study and at work much time can also be wasted in fruitless activity. When seeking information, try to find just the information you need as quickly as possible. When word processing, take care at all stages in the preparation of a document – but recognise when it will serve its purpose and the job is done.

7   As an aid to concentration, work to a job list (see page 16) and organise your work so that you engage in different activities. In particular, it is not a good idea to sit still – staring at a screen – for long periods. Take a break of at least five minutes every hour, exercising, relaxing, or working in a different way. This will help you to concentrate and will reduce fatigue.

Although you may be able to make more use of your computer to help you with your writing, you are advised to organise your work so that you spend no more time than you have to actually sitting and looking at a computer screen.

Making more use of your computer

Many who use a computer for word processing, for sending and receiving e-mail, and for obtaining information via the Internet, do not appreciate how they can use it in other ways to help them with their writing – with software programs that may already be installed in their computers. Although a program was developed to help users perform a particular task (for example, word processing), it may be installed as part of a suite containing other programs developed to help users with other tasks (for example, with drawing diagrams and charts, with desk top publishing, with preparing and delivering presentations, and with preparing and using spreadsheets and databases); and each of these programs may have capabilities that overlap with those of the others.

Desk top publishing

With desk top publishing software, page layouts can be planned in a choice of formats, with tables and figures in appropriate places close to relevant text. The result should be a finished appearance indistinguishable from pages in a printed newsletter, magazine, book, or other publication. With improvements in word processing software, however, the line between word processing (with a word processing program) and desk top publishing (with a desk top publishing program) is increasingly difficult to draw, and anyone considering preparing camera-ready copy for a publisher should ascertain the publisher's requirements before starting to write.

Preparing presentations

With appropriate software it is easy to prepare: a topic outline for a talk, speaker's notes, visual aids for use during a talk as overhead projector transparencies or as slides (see page 171), and handouts providing further details – for distribution after a talk. Slides (images stored electronically on a disk) can be prepared with or without a background colour and design; and both visual aids and handouts can include words alone, tables, charts, or other artwork – including photographs. However, care should be taken that the choice of background (see also page 169), or the use of special effects, is not such as to distract listeners – who should be concentrating on your message.

Using spreadsheets

In a spreadsheet data are entered in a table in which vertical ruled lines between the columns and horizontal ruled lines between the rows form a grid in which the resulting spaces are called cells. Whereas in a printed table, on a page, the number of columns and rows is limited by the type size used and by page size, a spreadsheet can be much larger – according to your needs. You can store data in cells; and by entering appropriate formulae in other cells you can perform calculations, analyse numerical data, and obtain statistics, as with a calculator. Furthermore, data saved on a disk can be edited and if you need to change an entry or add data in extra cells, or even add or delete whole columns or rows of data, recalculations are completed almost immediately and automatically by the computer. You do not have to calculate or recalculate.

Spreadsheets can be used for keeping records of your personal finances, and in business, for example, for recording and analysing sales data, and for accounts. As in word processing, spreadsheets can be printed as hard copy, and if necessary can be incorporated in word-processed documents. Results of the analysis of data, recorded on spreadsheets, can also be used to produce graphs, histograms and charts, and these too can be incorporated in word-processed documents (or in the handouts and visual aids used in presentations).

Preparing and using a database

With appropriate software, a computer can be used to construct and maintain a database, which in business could be used, for example, in keeping up-to-date staff records, or for stock records and stock control.

In a database, instead of storing records in a filing cabinet or card index, data are recorded electronically in a table and stored in a computer. Advantages of an electronic database are that: (a) it occupies less space than would a filing cabinet or card index used to store the same information; (b) records can be sorted easily and quickly – and data extracted – according to one's immediate needs; (c) it is easy to add, correct and delete records to keep them up-to-date; and (e) records are not lost or incorrectly filed – and so unavailable – as a result of the carelessness of some users.

Some people confuse spreadsheets with databases, but there is very little overlap in their applications: a spreadsheet is not a database. When planning a database, as when preparing a table, before entering any data you must decide the column headings to be used. In a database these headings are called field names. They indicate the kinds of information to be kept – in relation, for example, to each person or each item listed in the first column (called the stub in a table). In staff records the column headings would include: Surname, First name, Employee ID (the primary key: a unique identifying alphanumeric reference), Date employed, Post held, Department, and Salary.

In a database it is easy to amend records: to add or delete horizontal rows of cells (for example, in one database – as staff join or leave the business, or in another – as new kinds of goods are added to stock or as other kinds are sold out and not replaced), and to change the information in any cell (as, for example, people are moved from one department to another, or as the stock of each product changes from day to day).

However, adding columns (fields) is impossible with some databases and can cause problems with others. So the specification for a database must be carefully considered; and a business practice or procedure may have to be better defined before the specification for the database can be written. Then the database must be prepared by someone who understands the uses of a database.

The kind of database described here, in which operations are performed in one table, is called a flat file system. In another kind of database, called a relational database system, different tables can be linked by common fields so that when changes are made in one they are also made automatically, at the same time, in the others. This saves space in the database, eliminates duplication of effort – in data collection and data entry – and so saves time, and ensures that everyone using the database has access to identical data.

Databases can be constructed so as to provide different levels of data protection. For example, in an organisation some people may be able to view a particular screen showing personnel data complete (including personal details) whereas other employees accessing the same screen would not see such sensitive information.

People with good software skills should be able to design and construct a simple database, using a desk top PC program, but server databases and mainframe databases are designed by specialist database engineers.

Purchasing a computer

Anyone selecting and purchasing a computer is likely to have conflicting requirements, so some requirements cannot be completely satisfied. For individuals owning personal computers and for employers, some conflicting requirements result from the increasing rate of technological change. For example, obsolescence may make it desirable to update software as soon as possible, but because of the costs involved in purchasing new software and in acquiring new skills it may be necessary to delay making changes. For the employer there is also the conflict between the cost of continuing to work to existing standards, using existing procedures and obsolete equipment, and the cost of introducing new standards, new procedures and new equipment.

In relation to both the cost of purchasing a computer system and the decision as to the best time to buy, one should also bear in mind that any computer or information technology equipment you are thinking of buying will cost less, or will be obsolete and replaced by a more powerful and cheaper system, if you wait. The longer you wait, the more you are likely to obtain for your money, or the less you are likely to have to pay for a particular package.

In particular if you are a student and are about to start a course in higher education, you are advised to wait until you know: (a) what facilities are available at the college or university where you are to continue your studies; and (b) what your requirements are likely to be on the course for which you have enrolled. Then (c), if you decide to buy a personal computer, you can ensure that it will satisfy your needs and will be compatible with your college software.

References

Almack, J. C. (1930) *Research and Thesis Writing*, Boston, MA: Houghton Mifflin Co.

Albutt, T. C. (1923) *Notes on the Composition of Scientific Papers*, 3rd edition, Macmillan: London.

Beveridge, W. I. B. (1968) *The Art of Scientific Investigation*, 3rd edition, London: Heinemann.

BSI (1977) *Recommendations for the presentation of tables, graphs and charts*, DD 52: 1977 (a Draft for Development), London: British Standards Institute.

DTI (1988) *Instructions for consumer products: guidelines for better instructions and safety information for consumer products*, Department of Trade and Industry, London: HMSO.

Evans, H. (1972) *Editing and Design: Book 1 Newsman's English*, London: Heinemann.

Flesch, R. F. (1962) *The Art of Plain Talk*, London and New York: Collier-Macmillan.

Flood, W. E. (1957) *The Problem of Vocabulary in the Popularisation of Science*, University of Birmingham; Edinburgh: Oliver and Boyd.

Fowler, H. F. (1974) *A Dictionary of Modern English Usage*, 2nd edition revised by E. Gowers, Oxford: Clarendon Press.

Gowers, E. (1986) *The Complete Plain Words*, 3rd edition revised by S. Greenbaum and J. Whitcut, London: HMSO.

Graves, R. and Hodge, A. (1947) *The Reader Over Your Shoulder: A Handbook for Writers of English Prose*, 2nd edition, London: Cape; New York: Macmillan.

Henn, T. R. (1960) *Science in Writing*, London: Harrap.

Jay, A. (1993) *Effective Presentations*, London: Pitman (for Institute of Management).

Kapp, R. O. (1973) *The Presentation of Technical Information*, 2nd edition revised by Alan Isaacs, London: Constable.

Kennedy, J. S. (1992) *The New Anthropomorphism*, Cambridge: Cambridge University Press.

Land, F. W. (1975) *The Language of Mathematics*, London: J. Murray.

McCartney, E. S. (1953) *Recurrent Maladies in Scholarly Writing*, Ann Arbor: University of Michigan Press.

Napley, D. (1975) *The Technique of Persuasion*, 2nd edition, London: Sweet & Maxwell.

Orwell, G. (1946) Politics and the English Language, *Horizon* No. 76 (April, 1946). Reprinted (1957) in *Selected Essays*, Harmondsworth, Penguin Books, 143–157.

Partridge, E. (1965) *Usage and Abusage: A Guide to Good English*, 8th edition, London: Hamish Hamilton; New York: British Book Centre.

Potter, S. (1966) *Our Language*, 2nd edition, Harmondsworth and New York: Penguin Books.

Pullin, J. (2001) If you have got something to say, then say it simply. *Professional Engineer*, **14** (14) 80.

Quiller-Couch, A. (1916) *On the Art of Writing*, Cambridge: Cambridge University Press.

Sampson, G. O. (1925) *English for the English*, 2nd edition, Cambridge: University Press.

Strong, L. A. G. (1951) *English for Pleasure*, Methuen: London.

Stunk, W. and White, E. B. (1999) *The Elements of Style*, 4th edition, Boston, MA: Allyn & Bacon.

Tichy, H. J. and Fourdrinier, S. (1988) *Effective Writing for Engineers – Managers – Scientists*, 2nd edition, New York and London: Wiley.

Vallins, G. H. (1964) *Good English: How to Write It*, London and Washington: André Deutsch and Academic Press.

Index